CW00448239

Qu
Past and Present

by
Michael Roberts

Edited and Illustrated by
Sara Roadnight

Photographs by
Michael Roberts

Cover photograph:
Male and Female Coturnix Quail

Back illustration:
Painting of pair of wild Coturnix Quail

Published by Domestic Fowl Research
ISBN 0 947870 12 1

Printed by Bartlett & Son Printers
Swan Yard, Okehampton Street, St Thomas, Exeter.
Tel. 01392 254086

ACKNOWLEDGMENTS

Many thanks must be given to the following people and organisations for their assistance without which this book would have not been possible.

Major Henry Dumas
Rev. Brian Shillingford, Vicar of Kennerleigh
British Museum, Tring
Cairo Museum
Linnean Society
Dr. Kakizawa, Yamashino Institute
Victoria & Albert Museum
National Portrait Gallery
Julian Bird
Professor Dr. Mohammed Said Samy, Cairo
University of Agriculture, Cairo
and Edward Reeves and Jenny and Carmel Vella, for their kind and helpful information

INTRODUCTION

This book has been written from an English point of view, and certainly the historical side indicates this. Having said that, we have drawn heavily on French information and experience for modern production details. Many contacts have been made throughout the world, with field visits to Egypt and France.

During the research for this book, we were interested to note the low impact quail and quail products usually have in the market place, and certainly in England you have to ask if you want to see anything at all to do with quails, whereas in mainland Europe there seems to be a greater awareness of the product. Perhaps people in this country are wary of this unfamiliar product as they can be of certain exotic fruits and vegetables that appear in our supermarkets, and yet quail are quick and simple to cook.

One area of quail keeping that is very exiting is the new colours that are being created. I feel that in the next few years this will be of great interest as new colour variations come through and this might in turn lead to quail being shown.

MICHAEL D.L. ROBERTS
SARA ROADNIGHT
KENNERLEIGH 1999

4

QUAIL PAST AND PRESENT

			Page
1)	INTRODUCTION		4
2)	QUAIL IN THE WILD		7
3)	MAN'S ASSOCIATION WITH QUAIL		11
4)	HUNTING QUAIL		19
5)	QUAIL CALLS		23
6)	HOW DO I START?		25
		HATCHING EGGS	
		YOUNG LAYING BIRDS	
7)	METHODS OF KEEPING QUAIL:		26
	HOBBYIST -	A COOP AND RUN IN THE GARDEN	26
		INDOOR RUN OR CAGE	27
		AVIARY. MOVING QUAI	28
	COMMERCIAL -	ON THE FLOOR	30
		IN CAGES	30
8)	FOOD AND FEEDING SYSTEMS		36
9)	WATER AND LOW PRESSURE DRINKER SYSTEMS		38
10)	BREEDING -	IN-BREEDING	40
		COCK TO HEN RATIO FOR MATING &	40
		FERTILITY	40
		HATCHABILITY	40
11)	BREEDING OR LAYING SHED -		41
		LIGHTING	42
		TERRITORY	42
12)	SEXING -	COLOUR	44
		SIZE	44
		FOAM BALLING	44
		VENT SHAPE	44
		CALL	44

13)	EGGS -	COLLECTION OF EGGS	46
		STORAGE OF EGGS	
14)	INCUBATION	CANDLING	48
15)	BROODERS	HOBBYIST AND SMALL COMMERCIAL	52
16)	**COMMERCIAL BROODING AND REARING**		57
17)	**KILLING, PLUCKING AND FINISHING (HOME CONSUMPTION)**		62
18)	**COMMERCIAL SLAUGHTERING**		64
19)	**DISEASES AND AILMENTS -**		68
		QUAILS ON THE FLOOR	68
		QUAIL IN CAGES	69
		QUAIL IN AVIARIES	70
		QUAIL IN THE SCHOOL	72
20)	**ECONOMICS**		73
21)	**MARKETING -** FRESH QUAIL EGGS, COOKED QUAIL EGGS, QUAIL FOR ZOO FOOD, REARING QUAIL FOR PET SHOPS, QUAIL FOR LABORATORIES, QUAIL FOR MEAT, QUAIL FOR ETHNIC MARKET		74
22)	**SCHOOL PROJECT**		78
23)	**QUAIL SAYINGS AND QUOTATIONS**		79
24)	**WEIGHTS -**	IMPERIAL TO METRIC	82
25)	**LIST OF SUPPLIERS**		83

QUAIL IN THE WILD (COTURNIX)

There are two types of quail, Old World Quail, Coturnix Coturnix and New World Quail (i.e. the Americas) which are called Odontophorine. The New World quail have a serrated lower beak or mandible, hence the "odonto" which comes from the Greek for teeth. This beak may have developed like this because of the type of seed or food that the quail eat over there.

The colour of wild quail is dark brown, with a pale brown streak over the eyes and another on top of the head, running down to the nape of the neck. The back of the bird is brown, ash and black and is streaked with yellowish brown, with light brown feather shafts. A dark line runs from each corner of the bill, forming a kind of gorget above the breast which is a pale rust colour, spotted with black and streaked with yellow. The tail consists of twelve feathers, barred like the wings. Belly and thighs are yellowish white and the legs are pale brown. The female is distinguished by not normally having a black band down the centre of the throat and has less vivid plumage than the male. The female weighs in at about 5oz (140gms) and the male is smaller at 4 3/4ozs about (130gms).

Having given you this description of wild quail, it is quite interesting to note the enormous variation in colour and size of these birds as seen at the Natural History Museum at Tring in Hertfordshire. See photograph.

Coturnix.
The Quail

Coturnix fœmina.
A Hen Quail.

From Willughbys Ornithology 1678

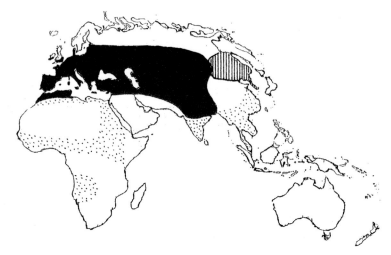

The map above is a rough guide to the distribution of the European coturnix quail. The dark areas are the summer migration range, and the dotted areas are the winter range. You can see the importance of Egypt as a migration corridor where most of the netting was carried out. The striped area represents the Asian coturnix quail summer migration range, with the winter range in South East Asia. There are other sub-species of coturnix quail in the Azores, Canaries, Madeira, South Africa and Madagascar, but the areas of their range have not been included.

The coturnix quail habitat is vast, spreading from Ireland and all of Europe across to the Ural Mountains, the Middle East, Iran, Afghanistan, Pakistan and India, and North and Central Africa (see map). There are several sub-species whose markings are different and these are to be found on the Atlantic Islands of the Azores, Canaries and Madeira, with two more sub species around South Africa. Coturnix Coturnix Japonica covers Korea, China, Japan, Mongolia and parts of Siberia, and it is from this species that the commercial quail of today has been created and improved.

European quail are summer visitors to the British Isles, arriving in April/May and leaving September/October; some have been known to over winter here as well. Quail used to be quite widespread in the British Isles with reports coming from most parts of Ireland, the Hebrides, Scotland, the Orkneys and the Shetlands, but this has all changed now. Although quail have never been very abundant in the British Isles, there were occasionally 'quail years' when more of them were seen than in other years. This was probably due to climate conditions, with France heating up very early, and more quail flying north in search of food. However, looking through the records, there seems to have been a marked decline in England since the 1750s, due in part to climatic changes.

8

Wild quail have a varied diet of grubs, insects, worms, larvae, grain and the seeds of grass and weeds. The male birds arrive first and set up territories which they guard jealously: there are reports of male quail fighting vigorously over patches of land. During this time, the male calls day and night to attract the in-coming females - quail do most of their migratory flying by night. It is said that one cock will attract several females onto his patch, but reports about the wild quail's social activities are sketchy and often conflicting, mainly because so little is known about this shy bird. The hens lay between 8-12 eggs (as many as 18 have been recorded) and sit very tight on the nest - it is possible to stroke a bird while she is sitting. Incubation time is 17-18 days. Sometimes she will rear two broods, one in June and one in August.

Once the chicks are hatched the cock bird takes no part in rearing them. They are the size of bumble bees, and will leave the nest immediately. At 11 days they can flutter and at 19 days they can fly. Unlike poultry, the young chicks scatter when disturbed, rather than cluster around the hen bird.

There are two interesting aspects to wild quail which are worth noting. Just before hatching quail chicks 'talk' to each other in the shell, in order to synchronise their hatching time, and this trait can still be found in commercial quail. The second point is that it is possible for early hatched quail to breed and nest in their first year, possibly in August, as hand reared wild quail have been known to lay at 32 days and the young males to be sexually active at 30 days. This might explain the over-wintering of some birds. One trait we do know is that quail are often very hard to flush, and when they do rise, they all fly in different directions, unlike partridge which rise up singly or in a covey but all in the same direction. The reason why quail fly in different directions is to protect themselves from predators of which they have many, particularly Lanner Falcons in the Sahara desert.

The European quail is the only gamebird to migrate. As mentioned before, they fly mainly by night, taking advantage of the right wind conditions. This is sometimes hazardous as they can become confused by lighthouse beams or drowned at sea. Major Partridge (yes its true!) reported in the autumn of 1925 "after a short spell of sea fog, ships off the Egyptian coast sailed for an hour through floating masses of drowned quail in at least one and probably two different localities".

Today, your best opportunity to see wild quail is during the summer months. They are shy birds, preferring the rough chalky grassland of the southern counties of England, particularly Wiltshire and Dorset, where they are admirably camoflagued by their primitive striped plumage. They also inhabit agricultural land as long as the crop is not too tall, but do not like woodland or overgrown vegetation. You will probably hear the male birds first with their trisyllable call, and may well trip over them before you realise what they are, as they sit very tight to the ground. If they rise up, they do so with a whir of wings, rocketing into the air and flying and gliding for a short distance, perhaps 100-200 metres, before diving for cover. They look a little like large brown starlings in flight but with slightly longer wings. Local ornithological groups would be able to help you to see them, otherwise quail are more plentiful in France and Spain.

It is useful to know the sizes of quail and partridge in order not to confuse them: a partridge is about 12 inches (30cms) long and a quail is about 7 inches (18cms) long from beak to tail tip.

It is just possible that, with the 'warming up' of our climate and the increase in Set Aside and SSSIs (sites of special scientific interest), we may gradually see more quail. One interesting fact is that my friends in Malta confirm an increase in migrating quail which is good to know, but how these birds might fare in Britain's modern countryside is anyone's guess, particularly as the English or Grey partridge is fading. Also two other problems await our intrepid quail, the greatly increased population of foxes which have affected so many ground nesting birds like the curlew and lapwing, and the disturbance caused by humans with their "right to roam".

QUAIL.—*Coturnix commúnis.*

MAN'S ASSOCIATION WITH QUAIL

The nature of this bird has led to easy captivity. Being a shy, gregarious and grain eating bird, domestication was not difficult, although it must be said that even commercial birds still retain many wild instincts.

Early remains of quail bones left over from hunter gatherers, are to be found in various caves around England and France. Two examples are Cow Cavern at Chudleigh and Kents Cavern near Torquay in Devon. These caves were explored extensively by William Pengelly and others, from Victorian times to the present day. Quail bones and other bird bones were found, along with the much larger bones of woolly mammoth, bear, hyena , lion and wolf, dating as far back as 80,000 BC; so quail were around then and were probably part of our ancestors' diet.

It was not until Ancient Egyptian times that quail were mentioned and depicted on a grand scale, not only on papyrus and bas-reliefs, but also in hieroglyphs, the figure of a quail denoting the letter W. Many people have mistaken this symbol for a chick or young chicken, which were unknown in Ancient Egypt at the time when this form of writing was invented. Chickens arrived in Egypt from Mesopotamia (Iraq) about 1450 BC. The use of a quail in hieroglyphs can be seen from the Early Dynastic period (2650-3000 BC) but it can be properly identified as a quail in hieroglyphs from 2575 BC onwards.

In a work routine written on a papyrus from Abydos it states:

> 1st day - Bread making
> 2nd day - Quail catching
> 3rd day - Ploughing
> 4th day - Sowing
> 5th day - Harvesting
> 6th day - Threshing
> 7th day - Relaxing and playing music

There is an account of a funerary report from Saqqara dating from the second Dynasty (1640 - 1532BC) which included "pigeon stew, cooked quails in addition to bread, porridge, fruit and kidneys, ribs and legs of beef".

Because quail were migratory birds in Egypt, and came in such large numbers, many of them would have been plucked, salted and dried in the wind. There are numerous illustrations of this practice, mainly depicting ducks and geese, but from the Roman Hipparelius we have "I like not the life which Egyptians lead, forever plucking quail and slimy magpies". Herodotus remarked that in Egypt, "quails and ducks and small birds are salted and eaten raw".

A drawing of a painting in the tomb of Nakht, Thebes, showing ducks being plucked, drawn and hung up to dry, before being placed in large jars. 1550-1307 B.C. (see passage of the Israelites in the desert).

About 1400 BC, we find Moses in the desert with the Israelites, clamouring for flesh to eat: Numbers 11, 31 "And there went forth a wind from the Lord and brought quails from the sea and let them fall by the camp, as it were a days journey on this side and as it were a days journey on the other side and as it were two cubits high upon the face of the earth". (A cubit is about 18" -20" long based on the length of a man's forearm). The last part of this piece was describing the size of the flock of quail as they arrived around the Israelites. The ornithologist Meinertzhagen, writing early this century, points out that quail fly low across the desert, about 36" from ground level (which corresponds to two cubits), and he goes on to say "Anyone who has watched the old gentlemen in the streets of Port Said during the second week of September, sipping their coffee and catching the quail in a butterfly net as they pass down the streets at dawn can testify that they fly 3' to 4' above the ground".

Continuing from the Bible: Numbers 11, 32, "And the people stood up all that day and all that night and all the next day, and they gathered the quails: he that gathered least gathered ten homers".

A homer is also a chomer from the Hebrew word meaning a heap. It was an old Hebrew measurement of capacity, also called a cor, containing 10 ephahs or 10 baths (a liquid measure) about 80 gallons or 360 litres. Some books have this as a dry measure of 48 gallons or 220 litres, roughly the same size as an oil drum (45 gallons), so you can imagine the number of quail collected!

From the German ornithologist Von Heuglin we have this observation, "In August, September and October, quail arrive in enormous numbers. During the journey across the sea which takes place with a north wind, the quail fly high and fast. On the dunes between Alexandria and the Nile delta, we often had the opportunity of watching their arrival. Dropping out of the skies almost perpendicularly, they come tumbling on to the sand, to the right and the left by the dozens and at such a rate, despite their proximity to each other, their contact with the ground is so violent and they are so exhausted that many allow themselves to be taken in the hand".

To continue with Numbers: "And they spread them all abroad for themselves round about the camp. And while the flesh was yet between their teeth, ere it was chewed, the wrath of the Lord was kindled against the people and the Lord smote the people with a very great plague".

A drawing of a relief of an offering bearer carrying a bunch of papyrus flowers, a bowl heaped with grain or a loaf, and a sheaf of corn with two quail strapped to it by their legs. From the tomb of Ramose at Thebes 1550-1307 B.C.

Most events in the Bible have a logical explanation but I was prepared to leave this as an act of god: perhaps they ate too much, or the birds hadn't been properly cooked, or even, God was just plain fed up with the Israelites! But I came across a report written by a Frenchman called Sargent concerning this phenomenon in Algeria which I will mention in a moment. I also began to realise that the Romans seemed to know a thing or two about the eating habits of quail because Pliny mentions, "Quails are very fond of eating poison seed, on account of which our tables have condemned them" They knew that quail ate hellibore, aconite, and hemlock without being adversely affected by them, but the meat of the bird was dangerous to eat, giving rise to death, partial paralysis, diarrhoea, vomiting, etc. From the French report by Sargent it was noted that these plants seemed to be particularly poisonous in the spring, a fact known to the Berbers in North Africa, so it is more than possible that the Israelites died from hellibore or hemlock poisoning transmitted by the quail; but before I start an 'Edwina Currie' scare, may I point out that quail meat in the shops today is not from wild birds but is all reared on quail farms!

The early Greek period gives us several insights into quail. Aristotle wrote, "The quails when they commence their flight if the weather is fine and the wind in the north, go into pairs and have a successful voyage. If the wind is south, it goes hard with them, for their flight is slow and this wind is moist and heavy. Those that hunt them, therefore persue them when the wind is in the south, but not in fine weather. They fly badly on account of their weight, for their body is large. They therefore make a noise as they fly for it is a toil to them".

Pliny then gives us a little more insight into the life of the quail, "The quails always actually arrive before the crane, though the quail is a small bird and when it has come to us remains on the ground, more than it soars aloft; but they too get here by flying in the same way as cranes, not without danger to seafarers when they come near to land: for they often perch on the sails and they always do this at night and sink the vessels. Their route follows definite resting places. They do not fly in a south wind, doubtless because it is damp and rather heavy, yet they desire to be carried by the breeze, because of their weight in their bodies and their small strength. This is the reason for that mournful cry they give while flying, which is wrung from them by fatigue; consequently they fly mostly in a north wind, a land rail leading the way".

There are four interesting points arising from both these accounts of quail. First, both Aristotle and Pliny only mention the south bound or autumn migration; did the quail in their day have another route for the northern spring migration? Secondly, the help given to quails by the north winds. It is known that quails fly-glide on migration hence their long wings, but in general they dislike flying which leads to the third point: I have not been able to find any other reference to the 'mournful cry' they give while flying. The fourth point about the land rail leading the way is interesting, as quail and land rails are two of the most unstudied birds in Europe, and it is fascinating to learn that they migrate together.

As we have seen earlier, neither the Greeks nor the Romans ate quail, preferring, in fact, to use them for fighting. Quail did not come back on European tables until Merovingian times, 500-751 AD, after which they began to enjoy gourmet status.

By the time the Romans invaded England, the Ancient Britons were already trapping quail, which they called sofliars.

During his travels Marco Polo, (1254-1324) describes seeing quail in eastern Iran (Persia) south of Yazd, on the road to Kerman, "Along the route there are many fine groves of date palms, which are pleasant to ride through and abundance of wild game including partridge and quail". Further on in his journeys he mentions two brothers, Bayan and Mingan who "are bound by covenant to provide the Great Khan's court everyday, beginning in October and continuing to the end of March with a thousand head of game, including both beasts and birds except quails". Quails would have migrated south from this part of central Asia at that time of year. While visiting the city of Kinsai (now Hangzhou) in China in the province of Zhejiang on the east coast, Marco Polo sees the "abundance of victuals, both wild game, such as roebuck, stags, harts, hares and rabbits, and of fowls, such as partridge, pheasants, francolins, quails, hens, capons and as many ducks and geese as can be told". When he reaches India he notes "indeed throughout India, the beasts and the birds are very different from ours (meaning Italy), all except one bird and that is the quail. The quails here are certainly like ours but all the rest are very different".

In 1384 Richard II gave to the Dean of St. Martin le Grand, "on account of the affection we bear to him" a very comprehensive permit for life, "............ to catch and kill as he knows best pheasants, partridges, plovers, quails, larks".

Aldrovandi (1522-1605) talks of the 'ortygokopos' or man "who offers these birds (i.e. quails) for sale or rather for sport. For he is called ortygokopos who carries a quail on his finger to the sport of quail-striking". Ortiki is the Greek word for quail.

These little birds were used for fighting by the Greeks, the Romans and the Chinese, a practice which is still carried on in Pakistan and India today, although artificial spurs are no longer fixed on to the birds' legs. Their owners, who often chew betel-nut, will rub their red saliva over the quail to give them a reddish colour as they believe this will make their birds braver. Two quails are placed in a small ring and a little corn or seed is sprinkled in front of them which induces them to fight. Quail tapping is slightly different and is practiced with one bird which is placed on the ground and tapped on the head. If it moves, the bet is lost for the owner, if it stays still he wins. Many side bets were made as well.

Meanwhile, on the other side of the world the Japanese were rearing quails for their song from early times. We follow with a section taken from the book The Birds of Nippon written by Prince Taka Tsukasa in 1932.

"The Quail has been a very favourite cage-bird in Japan from ancient times, and was once in favour with the Japanese in the years of Keicho and Kan-ei (1596-1643 AD), and this pastime was again revived during the years of Meiwa and An-ei (1764-1780 AD). The main object of keeping the Quail is to hear its calls. Its song is quite different from that of Bush-warbler or Skylark; it is rustic and sturdy. It is that rusticity and sturdiness that has gained for it the favour among classes of people such as the tea-virtuoso and samurai.

When it was in vogue during the Meiwa and An-ei, it had also come into favour with the daimyos. The 'Kiyu Shoran' (A book on Amusements and Entertainments) gives the following account of the state of things prevailing at that period.

"The cages are usually made in pairs; some are made of very expensive imported wood, inlaid with gold and silver; some are inlaid with mother of pearl and ivory, whilst others are finished in gilt lacquer. The Quail Show is held at regular intervals, and the hall in which the competition takes place consists of finely decorated tents made of costly woven fabric, gold brocades, and the stands on which the cages are exhibited are covered with the most expensive scarlet rugs. On the day of the show, all bird-fanciers, in Yedo, now Tokyo, participate in the competition, each competitor brings in his best bird. The competition is almost always held early in the morning, as the Quail sings best then. All the dealers in birds and bird-fanciers, of Yedo, who have asssembled in the hall hear the calls of the birds, judge them and render the decision, as the first, the second and so on. The official ranking of the birds are divided into two sides, namely, East and West, as in the official list of Japanese wrestlers and are written down on a large sheet of paper made of several sheets of expensive thick fluffy paper for ceremonial use, joined crosswise, and, the lists are posted on the east and west sides of the wall of the hall. The owners of the winning birds give presents to all the dealers in celebration of the victory. It is said that a considerable amount of money is expended on the occasion."

It is also written in the same book, that the Quail, unlike the Canary Bird, could not be taught to sing by a good singer, consequently, the owner of a good singer is very proud of the acquisition. As the Japanese kept Quails with the main object of hearing the good song, and the good singer is thus prized by them, they naturally begin to favour one particular style of song.

It appears in the 'Wakan Sansai Zue' (An Illustration of the Natural History of China and Japan) that the song of the wild bird sounds like "Chiji-kwai," but a bird that sings this note is rarely found among the domestic Quails of today, and, moreover, this note is out of favour now. There are birds that sing like "Cho-kichi-koh," "Kokkin-kwai," "Kee-hee-kwai," "Chokkwai," but neither of them are considered good. The best style of song is said to be "Kwac-kwac-kwai." It must be sung with a loud tone, sustained with a good long roll, and ended with a full and round voice. It is very difficult, indeed!

Subsequently, the keeping of Quails grew in favour from time to time, and when it came into considerable vogue about 1918-1920, a bird with a good song was said to be worth more than £100. During this period, not only the best singer was prized, but the eggs were also valued. In comparing the eggs before and after the fashion, it was noticed that there was a great improvement made in the size and weight of the eggs, I believe, therefore, that if the improvement of the egg was carried out scientifically, we should be able to have another interesting producer of eggs added to our poultry.

"Hence the song of the domestic Quail has become very different from that of the wild quail. The domestic birds sing 'Qua grrrr,' instead of 'Qua kah,' the usual wild song. The Japanese praise this style of song, and it reminds one of the sound of distant thunder, and high prices are asked for birds which prolong the sound of 'grrr' longest. Formerly the 'Quail Show' was held early in the morning as referred to above, but, at present, many people disregarded the time, for the Quails sing quite freely at all hours while on show, performing repeatedly against each other. During the recent years an added interest in quail keeping, is that the hen lays at least 150 to 250 eggs in one year, and the laying season can be arranged at the will of the keeper. A further interest lies in getting the Quail to lay in small cages. The cage is made of bamboo bars, and measures four-fifths of a foot square, the top being netted over and the floor strewn with sand, as the Quail are addicted to sand-baths".

It was from these song quail that the domestic quail of today is derived. Birds were bred for size and egg production and this in turn became a thriving industry in Japan, with about 2 million birds by 1940. All this was upset by the Second World War, luckily not before a number of exports were made to the U.S.A., but the original blood lines of the song quail were thought to be lost forever.

In Japan today they recognise three different strains of quail: wild, commercial and laboratory quail. The latter has its advantages in the lab, being small in size, having a relatively short commercial life (3 years, less in laying cages, 8 months) being resistant to disease, producing large numbers of eggs, being very light sensitive, and having a good metabolism.

Gilbert White (1720-1793) in his wonderful book the Natural History of Selborne mentions that "There are few quails, because they more affect open fields than enclosures; after harvest some few land-rails are seen" (remember Pliny). Later he states "Quail crowd to our southern coasts and are often killed in numbers by people that go on purpose".

Thomas Pennant, (1760) who wrote one of the first bird books states "Quails are found in the parts of Great Britain; but not in any quantity, they are birds of passage; some entirely quitting our island, others shifting their quarters".

Frank Buckland (1826-1880), the very interesting Victorian who ate or tasted most of the fauna he came across, including a 12th century saint's remains in a crypt in France, wrote "Quails are said to accomplish a hundred and fifty miles in a night and undigested African seeds and plants have been found in the crops of these birds when they reach the French coast".

Baron de Tott writes that "No country abounds in Quails more than the Crimea. During the summer they are dispersed over the country; but at the approach of autumn, they assemble together, and cross the Black Sea to the southern coasts, when they afterwards transport themselves to a warmer climate. The order of their migration is invariable. Towards the end of August, on a serene day, when the wind blows from the north at sunset and promises a fine night, they repair to the strand and take their departure at six or seven in the evening, and have finished a journey of fifty leagues, (150 miles or 240 kms) by break of day".

One final intriguing note comes from a Mons. Pellicot who wrote, "I saw some boats come in containing ten or a dozen sharks. They were all opened before me, and there was not one which had not from eight to ten quails in its body".

I have tried to introduce as many interesting vignettes as possible without, I hope becoming too repetitive, if only to give us a little more insight into the habits of quail as remarked on by man.

HUNTING QUAIL

As we have seen earlier in this book, wild quail often present themselves as being one of the easiest of game birds to shoot or net because of their incredible reluctance to fly, and preference for hiding in the undergrowth. (A group of quail is called a bevy).

Our first record of this is to be found in the bas-reliefs in Egypt of quail being netted. It appears that four or more men went out into the field with a rectangular net and made their way to a spot near the quail; (probably the Egyptians had heard them calling). The net was placed over the corn and held taut while two or more other men drove the birds towards it. In attempting to escape, they rocketed upwards into the net and were caught. Looking at these bas-reliefs in Saqqara, I had a feeling this was carried out possibly at night or early in the morning, as it appeared that some of the netters were carrying torches and/or rattles.

In 1751, the Swedish naturalist, Frederick Hasselquist noted, 'an amazing number of these birds come to Egypt at this time (March), for in this month the wheat ripens. They conceal themselves in the corn, but the Egyptians know extremely well that there are thieves in their grounds; and when they imagine the field to be full of them, they spread a net over the corn and surround the field, at the same time making a noise, by which the birds are frightened and endeavour to rise, and are caught in the net in great numbers and make a most delicate and agreeable dish'.

A drawing of a relief from the mastaba of Mereruka, Saqqara, depicting quail being netted in a stubble field - 2323-2150 B.C.

The other method of catching quail using vertical nets, seems not to have been known in very early times, but was used extensively along the northern shores of Egypt according to Diodorus (500BC). The use of vertical nets was started by fishermen who observed the flights of migratory quail and the fact that they became entangled in fishing nets which were hung out to dry. Knowing the migratory times of year, the height that quails flew and using what they had to hand, the fishermen were able to supplement their diets and their incomes.

Willughby, 1678, mentions 'For catching of quails they use this art: the fowler betimes in the morning having spread his net hides himself under it among the corn: then calls with his quail pipe. The cock quail thinking it to be the note of the hen that he hears, comes in a trice with all speed to the place whence the noise comes. When the bird is got under the net, up rises the fowler and shows himself to him, he presently attempting to fly away, is entangled in the net and is taken'.

Another method of taking quail was stalking with a gun. The hunter noted the wherabouts of the bevy and walked round them in a large circle to gather in any stragglers. He repeated this several times, the circles becoming smaller, until the bevy was quite closely packed and this method assured him of a right and left shot when the birds rose into the air. In the 1700's, the hunter would have shot them on the ground, but later there came a period in shot-gun development and sport which called for the hunter (gun) to shoot the birds in the air. There is a reference to a man in Lincolnshire in 1838, a good year for quail, who killed 16 birds with one shot, although these were presumably on the ground.

It is interesting to look through Colonel Peter Hawker's diaries; he was a keen sportsman and observer of nature who kept a record of every day's shooting, most of it in Hampshire, and in his lifetime he only shot 58 quails. This he recorded during 51 shooting seasons, and he died in 1853. Following from this there seem to be very few sporting prints of quails being shot; the only ones I have seen were French.

Robert Blakey writing in 1887 stated that "Quail shooting is chiefly confined in Britain to some sections of it: to the counties of Essex, Kent, Cambridgeshire, Suffolk and Norfolk," and Douglas Dewar writing of that period, noted "The Isle of Thanet was famous for its quails and people used to assemble there from afar to shoot them."

I remember rearing 50 quails in the late 1950's on our shoot at Dorsington in Warwickshire. When flushed they flew as individuals in different directions and could never be held to any particular piece of ground. Sometimes they rose with partridge, and being smaller, were taken for 'cheepers' or very young partridge, and were not shot. On the whole they were not a success. They were a fad in those days, much as American pheasants are today, but quails have been protected since 1981.

Continental sportsmen in general, are not keen on quails as sporting birds, because they often rise up at head height, which has caused a few fatalities and scores of people to be blinded by shot gun pellets.

In the Mediterranean countries quail used to be netted in grotesque quantities. Temminck (1778-1858) reports that over 100,000 quail were caught in a single day along the coast near Naples, and on the tiny island of Capri, quail were taken in such large numbers from the tenth century onwards, that the church extracted a revenue from them; hence the local inhabitants could afford a bishop who was called the Quail Bishop.

A drawing of long nets employed to catch quail near to the sea coast.

21

Howard Irby (1800's) states that "vast quantities of quail were caught in southern Spain in the Spring during the months of March and April, with small nets by the aid of the 'quail call'." and in Malta quail were netted or shot in large quantities during the migration months, April and September. In April, the cock birds were trapped by using a female caller, and then put in small cages to call in the passing females. These would be gathered in under the nets, in much the same way that the ancient Egyptians used to do it. The quails were loathe to take to the wing again, having flown across the sea from North Africa, so they would be herded in under the nets by men using long lengths of string with tin cans tied at intervals along them.

A more recent example of the mass slaughter of quail took place in 1908, when 1,000,208 quails were exported from Egypt alone for sale mainly to France, England and Germany. The birds were exported live, (as there was no refrigeration in those days), in low bamboo crates, to prevent head damage. Wild quail survive being caged quite successfully, but even so there must have been losses. By the 1920's, the number of migrating quail had dropped alarmingly and in 1937 Egypt and most of the Mediterranean countries signed an agreement to ban the taking of quail. Even so, netting continues today in certain southern European countries, where old customs die hard.

QUAIL CALLS

Most people associate coturnix quail with the shrill trisyllable call of the male. Both the male and the female have a series of quiet calls, and it is only by keeping them near at hand and listening that you will be able to hear them.

I have not been able to find the origin of 'wet my lips', which the male call is commonly known as. It actually sounds nothing like that, more 'whit-ti-tit'. Most attempts by humans to put words to bird calls or sounds are reasonably successful, like 'pee-wit' for lapwing or peewit, 'chiff-chaff' for chiff-chaffs and 'a little bit of bread and no cheese' for yellow hammers. I can only assume that the origin of 'wet my lips' is derived from tired and thirsty agricultural workers toiling in the fields: in years gone by, when there was little or no machinery, work in the fields was long, hard and thirsty, and part of the perks of the job, was the provision of flagons of cider and ale, so wetting the lips was a regular occurance during the day.

There are various regional variations of this name such as 'wet my feet', and from Cheshire 'but for but'. Onomatopoeic names for quail include two French ones: 'j'ai' du ble' and 'jai pas de ca' (the de is not pronounced). The Germans have 'Buck-den-ruch' and there are probably countless others.

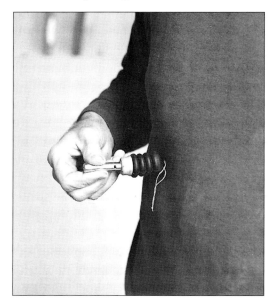

A modern quail pipe, which is pressed against your flank and imitates the call of a female quail. This is for calling male quails under nets.

23

A portrait of Charles I 1600 - 1649, wearing quail pipe boots. By courtesy of the National Portrait Gallery.

I have so far been unable to find any quail pipes and can only think that they were made of clay or wood and have not survived. I imagine that they may have had a pipe made from a short piece of hollow cane, clay or wood with a crinkled leather bellow to make it work. I did find a modern quail call in France (see illustration) but I fear that the male quail would have to be awfully hard pressed to imagine that the noise emitted by this call was a female quail! A fashionable boot was developed in the 16th century from quail calls and Chaucer (c 1400) in his Romaunt of the Rose mentions: "high shoes that are wrinkled like a quail pipe", while later in 1602 T. Middleton writes of "a gallant that hides his small-timbered legs with a quail pipe boot". The boot was mid calf height and made of soft leather which when worn, fell into wrinkles round the leg. They were considered the height of fashion, and it is a pity that neither they nor the quail calls seem to have survived.

A pillar photographed at Saqqara, near Cairo, depicting a quail which denotes the letter W.

A hieroglyph of a quail, obviously a young one, from the mastaba of Atet, Meidum. 2575 B.C. - 2465 B.C., Cairo Museum.

A relief from Saqqara showing cattle, fish, hippo and quail 2323 - 2150 B.C.

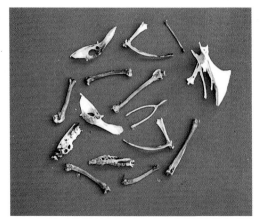

Bones like these have been found in caves around England (Chudleigh) and Europe.

Cow Caven, Chudleigh, Devon, where quail bones have been found dating back as far as 80,000 B.C.

These photographs show the varying colours to be found on wild quail, mainly collected in Victorian times, some by Arthur Wallace. By courtesy of the British Museum, Tring.

Quail eggs, the ones on the right will hatch better than the ones on the left which are slightly cloudy, possibly showing a hormonal imbalance.

A view of the largest quail organisation in France.

Laying cages in Cairo University. Note the whole system is on castor wheels.

Sara by a signpost of a village in France, La Calliere, which means, the place where quail are kept.

Quail eggs, showing the varying blotches and colours.

A foam ball attached to a dropping. Note its foamy texture.

A Brinsea Octagon 20 incubator with quail chicks just hatched.

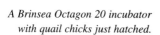

A Brinsea incubator with chicken and quail eggs to show the comparison in size.

HOW DO I START?

There are two ways of starting with quail: a) you buy hatching eggs, or b) you buy young birds just ready to lay or in lay.

a) **Hatching Eggs.** The normal hatch rate for commercial quail is about 90% - 95%, so out of a dozen eggs you should hatch eleven chicks. Assuming that there is approximately a 45/55 F/M sex split, a dozen eggs should produce 4 to 5 hens. Eggs should be bought direct from the breeder, and ideally should not be more than a week old, but you can stretch this to 10 days after which hatchability declines. I have known some people who have hatched eggs from the local health food shop, but I think they were lucky! If you buy direct from the breeder you know what you are getting and how fresh the eggs are. Now you need to turn to the chapter on incubation.

b) **Young laying birds.** Where do I buy them? There is a list at the back of the book of breeders both in England and the U.S.A., otherwise there are publications such as Country Smallholding, The Smallholder, (both monthly), and Cage and Aviary Birds (weekly) in the U.K., and Game Bird and Conservationist Gazette in the U.S.A. Addresses of these are also at the back of the book. How do I know the quails are young? Look at their feet and beaks - they should be clean with naturally short nails and no clipped beaks and the feet should not have any bunions. The birds should be plump, with even unbroken feathers, and the vent should be dry. It is unlikely that someone will sell you old birds, as quail reproduce so quickly. Quail will live for 3 - 4 years.

The next important question is, what type of coturnix quail do I want? There are four main classifications: pet or aviary quail (these are mainly coloured birds and are generally poor egg layers); dual purpose which are the most commonly found quail; Italian layers, these are small pale coloured quail, and meat quail (these are large quail, nearly the size of a partridge) usually poor egg layers and not readily found in England at the moment.

Many people in this country keep the dual purpose laying quail, as the meat market is still small here and is mainly supplied by imported quail from France. The dual purpose breed gives you the freedom to choose your markets: eggs, pets, some human consumption or zoological food. (This is a market where quails are sold dead, either fresh or frozen in the feather, for feeding to hawks, owls, eagles, snakes, reptiles or other zoo creatures).

At the moment and for the last few years there has been competition between quail breeders to produce larger and larger birds for the table. Because the coturnix quail has been domesticated for so long, it has a reasonably calm temperament, as against the wild coturnix or the American Bob White. Breeders are trying to increase body weight and reports indicate some new strains of over 1lb or 450gms in weight are being developed. Early indications show these birds to be slow in maturing, about 10 weeks before they lay, and poor egg layers too.

I have not mentioned buying day old chicks because most breeders do not sell them.

METHODS OF KEEPING QUAIL

As with all livestock, quail require proper housing, not converted rabbit hutches or the like. Quail are different from poultry and so are their requirements.

Where I housed my quail, there is an overhanging oak tree. In the autumn, the acorns dropped onto the tin roof, sounding like a pistol shot every time one fell. This is the sort of thing that stresses quails; they hate any sudden change, be it noise, movement, shadows, light etc. They will rocket into the air, or crowd into the corners at night time, so care should be taken over siting your quail building! If you live in an urban area, remember the male bird's call is quite shrill, so keep them away from nearby houses.

Having observed all the methods of keeping quail I must say I prefer the "deep litter" system of the small quail farmers, but sadly this is impratical on a large scale. Quail love to dust bath, and if you take a bird out of its cage and pop it into a box with some shavings, it sounds like a mini explosion within. One problem that arises from keeping quail this way, is what to call their eggs. They are not free range quail, as they don't go outside, and barn laid eggs have a poor image in the public eye.

A COOP AND RUN IN THE GARDEN

Quail in a coop and run in the garden.

26

Quail can be kept this way but there are several points to watch for.

a) Quail can withstand the cold quite well, but they can't stand wet and cold, and must have access to drinking water that is not frozen, therefore they should be brought into a shed or garage during the winter months.
b) They are susceptible to stress caused by cats and dogs and other things like rats, mink, fox, weasels etc.
c) They do not like sudden noise or commotion and will flutter up and bang their heads against the roof of the house or run.
d) They are more susceptible to diseases which can be transmitted through the soil and grass, or by wild birds.

I have been rather negative about keeping quail outside, although many people do this, but you need to be aware of the possible problems. The litter in the house can be straw or shavings, and you will need a small drinker and feeder. The 1 litre plastic type or cage cups are excellent. Remember to have a covered area in order to keep the quail food dry. Wet or damp quail food becomes mouldy in no time and is harmful to the birds.

Some people like to have quail running around the garden. This is possible but there are a few points to watch our for. Quail do not have a homing instinct, so you need to hold up the cock bird to call the hens back at night time. Watch out for cats and sparrow hawks; and lastly it is illegal to release Japanese quail into the wild, as they are a non-indigenous species.

INDOOR RUN OR CAGE

If you look at the photograph you will see the sort of small scale quail housing you can have in your shed at home or in the classroom at school. The sides need to be a little higher than those pictured as the quail love to dust-bath in the shavings which fly everywhere! This was a cheap construction which allowed good visibility and ease of management. It can be made in various modules, so that the

An indoor cage for quail, ideal for a school project. The sides could be 3" higher to stop the sawdust or shavings from scattering around - Size: 36"x18"x12" & 18"x18"x12"

quail can be moved easily from one to another at cleaning time (once a week), or selected birds can be housed separately to study egg patterns and colouring. The roof is made from a fine plastic netting, so if the birds do flutter up, they do not scalp themselves; it is also low enough - the dimensions are not too important but the height should be no more than a foot. The floor is the top of a table or bench, covered with newspapers and 1/2" of shavings. The birds are ushered into the small section and shut in, and the larger cage is lifted up, and the paper rolled up with the dirty shavings inside. The birds are then ushered back into the cage once it has been cleaned and provided with fresh litter, and the small section is then removed and cleaned. Two cage cups hang on the wire for water and food.

AVIARY - Coturnix quail are often used as 'cleaner uppers' in aviaries where there are paraqueets, canaries etc. They thrive on the bird seed which is wasted by the birds at the top of the flight. The quail just run about feeding on the floor of the aviary, but they do require a covered area that they can creep into where they will be safe from being "bombed" from above. This is in the form of a wide shelf, about a foot off the ground. Quail food and water can be placed under this shelf if required. The floor of the aviary should be well drained, i.e. raised and covered with pea gravel, and it should be cleaned and raked regularly. It should also be rat proof. The feeders and drinkers can be cage cups or 1 litre drinkers and feeders. Again, as with birds kept outside and on the ground, birds kept in aviaries are susceptible to various diseases.

One point to watch if you have security lighting around your aviary, is that it does not stress the birds at night time. Location and positioning are all important to avoid this.

MOVING QUAIL
Here are a few tips about moving quail.

a) If you keep quail already, it is best to isolate any new stock you buy for three weeks, just in case they bring in any infection with them.

b) Try to place new birds into their cage or aviary etc in the dark. Quail roost on the floor.

c) If you are mixing two batches of quails together, make sure they are all the same size, and always mix them in a new pen. There will be a little friction to start off with, but they should soon settle down.

d) If placing quail in an aviary, clip one wing so that they don't fly up and damage themselves on the roof of the aviary.

Top view of cage, without the lid. Note the sliding door between the two compartments, and the lid which lifts up and slides along.

Quail shut in the small compartment so the larger compartment can be cleaned. The top has a plastic mesh to stop any head damage.

e) Sometimes you will find a dominant bully hen. She can be a nuisance and it is best to remove her and/or cull her.

f) Never place young quail with old quail, or leave the male quail in the aviary with a female quail if she has young. He will normally kill the lot.

g) One method of taming your quail, having settled them into their new surroundings, is to feed them meal-worms. They will sell their souls for meal-worms, once they have a taste for them.

h) Always buy a little food from the person you have bought your quail from, to help minimise the stress of moving. Completely new food is stressful.

COMMERCIAL - There are two ways, indoors on the floor, or indoors in cages.

ON THE FLOOR - Small quail farmers use sectional areas approximately 6' x 4' (180 x 120 cms) with a door on one side. See illustrations. These are very manageable, as you can lean into the pen for cleaning, egg collection, feeding and watering without disturbing the birds too much. The concrete floor in the barn should be covered with wood shavings to a depth of about 1" - 2" (2.5 - 5.00 cms). You can keep about 60 adult birds (45 hens and 15 cocks) in a pen of this size. Clean it out approximately once a week with a dustpan and brush, dealing with the whole area section by section, moving the birds around as you go. Food should be given in a food hopper hanging from the enclosed roof; you can adjust the height to suit the type or age of quail. Water is normally given in plastic fonts set up on small bricks or slabs of stone so that they are above the shavings. Because of the quails' tendency to dust bath, the water font would fill up with shavings in no time if it were put straight on the floor.

IN CAGES - Apart from those reared on the floor, most adult quail are kept in cages. Nearly all the quail used for egg laying are caged, whether the eggs are for human consumption or for hatching. There are standard cages made, but I have seen many different home-made ones which seem to work, some in metal, some wood and metal, and mainly using 1" x 1/2" weld mesh (2.5 x 1.25 cms). Dimensions vary according to the breeder, but the height is never more than 8" or 20cm. There are two reasons for this: firstly, the quail have less room to jump and bang their heads, and secondly, more cages can be stacked up on top of each other. Most European cages are 4 or 5 layers high, but I believe the Japanese manage 7 layers high.

DETAIL DRAWINGS OF FRONT AND SIDE SECTIONS OF QUAIL LAYING PENS

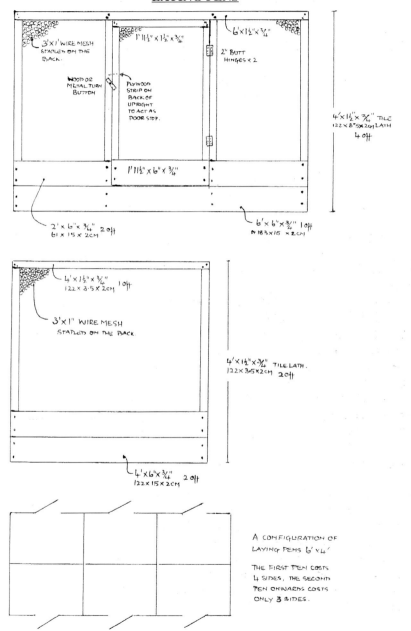

3' X 1' WIRE MESH STAPLED ON THE BACK.

1' 11½" X 1½" X ¾"

6' X 1½" X ¾"

2" BUTT HINGES X 2

WOOD OR METAL TURN BUTTON

PLYWOOD STRIP ON BACK OF UPRIGHT TO ACT AS DOOR STOP.

4' X 1½" X ¾" TILE 122 X 3·5 X 2CM LATH 4 OFF

1' 11½" X 6" X ¾"

2' X 6" X ¾" 2 OFF 61 X 15 X 2CM

6' X 6" X ¾" 1 OFF OR 183 X 15 X 2CM

4' X 1½" X ¾" 1 OFF 122 X 3·5 X 2CM

3' X 1" WIRE MESH STAPLED ON THE BACK.

4' X 1½" X ¾" TILE LATH. 122 X 3·5 X 2CM 2 OFF

4' X 6" X ¾" 2 OFF 122 X 15 X 2CM

A CONFIGURATION OF LAYING PENS 6' X 4'

THE FIRST PEN COSTS 4 SIDES, THE SECOND PEN ONWARDS COSTS ONLY 3 SIDES.

31

DETAIL DRAWING OF QUAIL LAYING CAGES

<u>TOP VIEW</u>

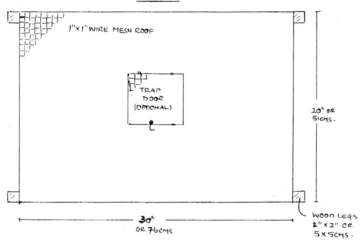

1"x1" WIRE MESH ROOF

TRAP DOOR (OPTIONAL)

20" OR 51CMS.

30" OR 76CMS

WOOD LEGS 2"x2" OR 5x5CMS.

<u>SIDE VIEW.</u>

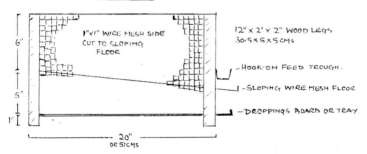

6'

1"x1" WIRE MESH SIDE CUT TO SLOPING FLOOR

5'

1"

20" OR 51CMS

12"x 2" x 2" WOOD LEGS 30.5x5x5 CMS

-HOOK-ON FEED TROUGH.

-SLOPING WIRE MESH FLOOR

-DROPPINGS BOARD OR TRAY.

THE SLOPING MESH FLOOR MUST BE ADJUSTED SO THE EGGS ROLL TO THE FRONT BUT DO NOT CRACK AGAINST THE FRONT STOP OR AGAINST EACH OTHER.

WATER DRINKERS ARE POSITIONED ON THE BACK OR THE SIDE OF THE CAGE, SOMETIMES **ALSO** ON THE FRONT

<u>FRONT VIEW</u>

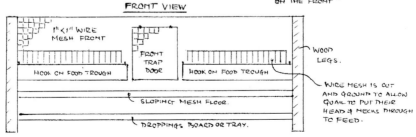

1"x1" WIRE MESH FRONT

FRONT TRAP DOOR

HOOK ON FOOD TROUGH

HOOK ON FOOD TROUGH

SLOPING MESH FLOOR.

DROPPINGS BOARD OR TRAY.

WOOD LEGS.

WIRE MESH IS CUT AND GROUND TO ALLOW QUAIL TO PUT THEIR HEAD & NECKS THROUGH TO FEED.

32

The illustration shows the layout of the cage. Allow 25sq inches per bird. The main points to remember when constructing cages are, ability to inspect the birds and ease of access for egg collection (most cages have roll away floors), ease of access to the birds for cleaning out, feeding and watering (water is nearly always supplied by low pressure systems,) and the birds must have sufficient access to artificial light.

Feeding is normally by hand with the aid of a square scoop. Egg collection is by hand into plastic buckets or straight onto egg trays.

The droppings are cleaned out in a variety of ways. Some cages have metal trays which are pulled out and scraped clean, some people use paper feed bags, others use a roll of paper under the cages, which is drawn out at the end. The most expensive cages have rubber conveyor belts under each tier or layer, which empty at the end of the cage stack.

Cleaning out is most important as quail manure has its own strong distinctive smell.

It is a good idea to fix castor wheels onto laying cages so that at the end of a laying cycle (8 months) the cages can be wheeled out to be power washed and disinfected.

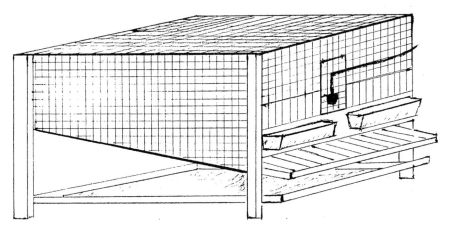

A typical laying cage, with automatic drinker, food troughs, roll away egg floor and droppings tray.

4 tier laying cages. In Japan they manage 7 tiers, but with greatly reduced head room for the quail. These cages are farm made with dexion angle iron and wire mesh.

A close up of the same cages. Note the drinkers: two per cage are on the front of the cage, and the automatic feed troughs are at the back. Access to the cage is from the front.

One end of a shed with laying cages in France. Note the plastic mouse trap on the floor.

A quail farm in France with 1/4 million quail.

35

FOOD & WATER

First of all, when planning to keep quail, you must know the correct amount of space required for feeding and watering your birds. If the space is insufficient birds will be bullied away from the feeders and drinkers and be non productive.

	CHICK	GROWER	LAYER/BREEDER
Feed Space	3/4" 2cm	1" 2.5cm	1 1/4" 3cm
Water Space	1/4" 1cm	1/2" 1.5cm	3/4" 2cm

The water space above is not very specific as nipple drinkers are used mostly today and you require 2 nipple drinkers to each cage

A feeder and a drinker (1 litre size) useful for the smaller quail farm or school project.

FOOD - One of the problems in the UK is finding someone who supplies proper food, because the quail industry is small and fragmented. There are mills who produce it, but whether they make it all the year round is another matter; you would have to telephone around the feed mills to find out. So how does the quail industry survive, particularly the small quail farmer? Partly on Turkey chick crumbs, most of which can be loaded with antibiotics (like dimetridazole, an anti-blackhead drug), partly on poultry food (which normally has too low a protein content) and partly on game food (partridge breeder) which tends to be rather seasonal. One problem with all made up animal food is that it has a shelf life of only 4 months, so it is no good ordering tons of food to last a year. Some quail farmers mix their own, so that they can have additive free food at the right protein level which they can adjust up or down seasonally. A typical house mix will include ground wheat and maize, soya bean, fishmeal, vitamins and minerals.

The protein requirement for chicks and growers up to the age of 4 weeks is 26% and for your adults and layers 22%.

GRIT - This is included in proper quail food and is normally powdered oyster shell and/or limestone. Remember quails are pumping out eggs and therefore require a higher calcium intake than other made up foods for chickens or turkeys would provide, so additional small bird (budgerigar) grit is very suitable.

WATER - As I have explained elsewhere, water is normally given in plastic chick drinkers or on a low pressure water system. Quail drink plenty of water, and a drop

in egg production nearly always means a fault in the water supply; 24 hours without water will stop the quail from laying for a week, and longer will put them into a stress moult, so careful monitoring of the drinking system is very important.

Most medicines, if used at all, are water based, which makes them very effective.

LOW PRESSURE DRINKER SYSTEM

This system is a little like 'Lego'; you buy piping, and plastic inserts in the form of 'T's', junctions, straight joiners and 90 degree bends plus your drinkers. There are two types of drinker, nipple or nipple with cup.

You require a header tank to begin with. Inside the tank is a ball or float valve, so that the water fills to the level required. The tank needs to be placed above the height of the cages or pens where the drinkers are to be used, and it needs a lid to keep the water free of dust, flies etc. The size can vary, but most people have a one gallon header tank.

A small hole is drilled to accomodate the low pressure pipe about 2" above the bottom of the tank. This helps prevent any foreign bodies from entering the low pressure system.

Fix an on-off valve just outside the tank on the low pressure piping, so if you want to drain the system down or you have a leak, this can be dealt with without draining the header tank.

From now on it is a matter of cutting the tube and pushing it onto the plastic inserts. Use a Stanley knife and cut squarely across the tube. Have a boiling kettle handy and dip the end of the tube into the hot water for 20 seconds, to soften it. That way it will fit easily onto the inserts.

When making up the system, try to cut the pieces the same length for the various sections so that when you come to clean it out, all the tubes are interchangeable. The other important point is to place on/off valves to each branch of the system so you can isolate a particular branch for cleaning, or repair etc, without shutting the whole system down. You can bend the pipe back on itself to stop the flow of water, but after a year or two it will crack or split. You will require pipe clips or ties to keep it tidy.

The two main problems with this system are airlocks and leaky drinkers, so always check that everything is working.

A low pressure water system showing a header tank,
float valve, piping, taps, tee pieces, clips and drinkers

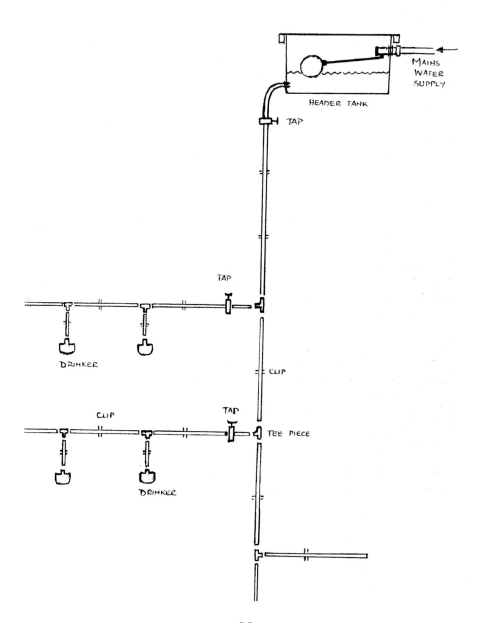

MAINS
WATER
SUPPLY

HEADER TANK

TAP

TAP

DRINKER

CLIP

CLIP

TAP

DRINKER

TEE PIECE

BREEDING

IN-BREEDING Most people are looking for 'hybrid-vigour' in their quail to maximise the number of eggs laid by each bird. It is possible to in-breed or line breed, in order to reproduce certain colours, but if you do this you may possibly see a drop in egg production, more infertile eggs, more dead in shell, and more deformities (this shows up mainly in the leg and foot department), although normally this is not a problem.

COCK TO HEN RATIO FOR MATING AND FERTILITY

The norm is 1 cock to 3 hens, but this can vary from season to season. Persistent bullying by the males can often cause quite serious injuries to the females. In smaller egg laying breeds the ratio can come up to 1:6, but in the heavier meat breeds the ratio is normally 1:3. You will need more males in the winter than in the summer as they will not mate readily when they are cold..

HATCHABILITY Most commercial quail farms have a hatching rate of 93-97%. This is why they do not bother to candle their eggs. The large meat quail are poorer layers with poorer hatchability at the moment.

Young quail in a sectional pen, nearly ready for the laying pen or cages.

40

BREEDING OR LAYING SHED

As with most farm buildings, access and location are very important factors. A garage could be a starting base, but the buildings need to be insulated, as when the quails are installed the inside temperature needs to be 70 degrees F. Conversely, extractor fan(s) are required during warmer months to reduce the inside temperature. Quail themselves produce a fair amount of body heat so heating inside should not be necessary, and in warmer climates I have seen air-conditioners attached to layers' sheds. Obviously you require electricity and water, and the building should be vermin proof.

Dust and smell are two other points to consider when quail are kept intensively. Dust is produced from the quails' feathers and the food, and as mentioned elsewhere, quail manure has a strongish smell.

Laying pens. Note the hanging food hoppers and the pigeon drinkers.

You must make provision for servicing your cages or pens when laying them out inside your shed. The construction can be of many types of materials. A smooth concrete floor is important for wheeling cages out to be power-washed, and also for cleaning. It is advisable to have a false ceiling, and this can be made of plastic sheeting.

Although your building may be well insulated, the sound of the male quail does carry, so remember this when positioning your laying shed. Egg production from the females increases by 10% to 15% without any cock birds in the pen. Males are only used to produce hatching eggs.

Access can be via two large doors for wheeling equipment in and out, with a service door at the side. Wire mesh doors can be fitted inside so on hot days the main doors can be opened and extra ventilation provided, without wild birds flying in or escapees getting out.

In general, nobody builds a quail breeding house as such; usually these houses or sheds evolve from other farm buildings, and very effective they are too.

LIGHTING

Laying quails need 14-16 hours of light. In commercial farms, caged quails don't see daylight, only fluorescent tube lighting which is hung between the cages so the bottom tiers benefit. These tubes are on timers, and the laying house has dim 'night' bulbs as well.

Daylight is used on smaller farms, but the length of day is adjusted in the early morning, so that lights come on to ensure that the birds have 14 hours of light a day.

At the end of the year and in early Spring, the lighting needs to be increased or decreased. Do not give the birds full 'lighting time' in one week, otherwise you will stress them and they will go into a moult. Adjust the lighting up or down in 1/2 hour amounts each week. The best system is to add the 'light 1/2 hours' in the morning so the quails have the natural dusk to settle down in for the night. The lighting does not have to be bright, 60 watt bulbs are ideal.

TERRITORY

Quails still retain territorial instincts even though they have been domesticated for centuries, so great care should be exercised when moving birds from cage to cage, or breeding pen to breeding pen. If you wish to add more birds to an established group, it is best to move the established birds and place them in a new pen or cage with the new birds, because otherwise there could be a blood bath; even adding extra hens to an established group can be fatal, so topping up a pen of birds is unwise. Set up your breeding pen and leave it as it is, unless you want to change all the cocks or all the hens.

A nice pen of laying quail on shavings. Note the 6" (15cms) kick board around the sides of the pen.

SEXING

SEXING - Quail are easy to sex if they are natural coloured and adult, see photograph, although confusion can arise when buying young birds and especially coloured varieties. Do be aware that there are sexless or hermaphrodite quails from time to time, but in a normal situation here is what to look for:

A) COLOUR - In natural coloured birds as seen on the front cover of this book, the difference in sex is easily seen. The males have a reddish brown breast and the females have a speckled breast. This is apparent when they are three weeks old, and is why most commercial farmers use only this colour and the lighter Italian laying quail, where the cock birds have reddish cheeks or faces.

B) SIZE - This is not apparent until the birds are at point of lay. The females are about 25% larger than the males. When you have been handling quails for a while, you can tell the sex of a bird by the size of the 'handful'.

C) FOAM BALLING - When the male bird is fully sexually mature (from 6-7 weeks old), this will be noticeable by the swollen cloaca gland under the tail. Willughby writing in 1678 noted that "the cock had great testicles for the bigness of its body, whence we may infer that it is a salacious bird". This is a common mistake, as this gland does not produce sperm, rather a urinate from the kidneys in the form of a sticky foam. If you gently squeeze the gland, the foam will squirt out, a little like shaving cream out of a can. The cock bird drops these foam balls as territory markers, and as far as I know is the only species to do so. At migrating time in the wild, this gland disappears, probably to help streamline the quail. The gland can be quite swollen and large for the size of bird, about as big as a small walnut. The size is also an indication of the birds' sexual maturity and dominance; some males loose this swollen gland if they are in a pen with too many other cock birds and are being bullied.

D) VENT SHAPE - This is an indication of whether the bird is in lay or not. A hen bird that is laying well has a wide horizontal shape to its vent, with a slight tinge of blue to the opening, while the non laying female or immature male has a small round vent.

E) CALL - This is more apparent when the male bird is fully mature; he emits a shrill trisyllable call.

Sexing quail. The female with the spots is on the left, the male on the right.

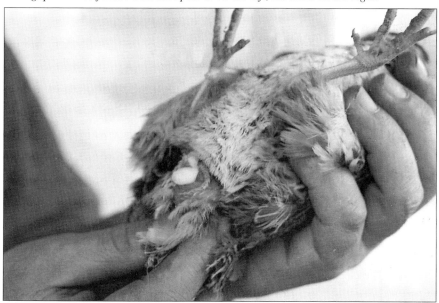

This shows the swollen male Cloacan gland being gently squeezed to produce a foam ball.

EGGS

COLLECTION OF EGGS - This should take place in the late afternoon or evening, as 75% of quails tend to lay between 5 pm and 7 pm. Two small plastic buckets or containers can be used, one for clean eggs and the other for dirty ones. Clean eggs can be put straight into the plastic egg trays, while dirty eggs can be washed in a sieve or chip fryer, but must be used just for hatching. Only clean eggs can be sold to the public. This is one of the problems of 'free range quail': unless the litter is changed very regularly, there will be a high percentage of dirty eggs.

STORAGE OF EGGS - Once placed point downwards in the trays, the eggs should be turned regularly. This is done by propping up the egg trays with a piece of wood on alternate sides, changing the position 3 times a day. Because the yolk is large as compared with the size of the egg, there is more chance of it 'sticking' to the side of the shell if not turned.

If you have only a small quantity of eggs, they can be placed in a seed tray filled with dry sand. Line them up all pointing in the same direction across the tray, and turn them end over end. This way you can check that they have all been turned. Look out for dents or cracks which are more apparent after 24 hours when the eggs have cooled down. Keep them out of direct sunlight and store them at a temperature of 16-17 degrees for a maximum of ten days. Some people don't bother to turn them during the first seven days.

In the commercial world, the quail eggs are collected in plastic buckets and then washed, checked, loaded into the incubator trays and placed in the incubator. The doors of the incubator are left open, and the turning mechanism switched on without the heaters. The eggs are stored and turned like this until required. Eating eggs are collected straight into the egg trays.

Eggs that are badly turned, stale, or badly stored often result in crippled chicks.

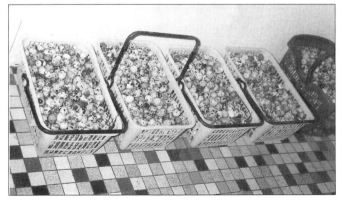

Quail eggs collected commercially and ready to be washed prior to going into the incubator.

A quail's egg weighs approximately 10gms (1/3 oz) as against a chicken's egg weighing 55gms (2oz) and is approximately 8% of the body weight of a quail, as against the chicken's egg which is 3% of a chicken's weight. In many ways a quail's egg is more nutrious than a hen's egg; here is a comparison table.

	Whole quail's egg	Whole hen's egg
Egg weight gms	10	55
Yolk %	29	25
Albumen % (white)	55	61
Calcium (mg)	59	57-61
Phosphorous (mg)	220	123 - 237
Iron (mg)	3.7	2.1 - 2.8
Thiamine (mg)	0.12	0.07 - 0.09
Riboflavin (mg)	0.86	0.27 - 0.32
Niacin (mg)	0.10	0.05 - 0.32
Energy (kcal)	158	155-183

Of course these figures will alter with the size of egg and the food and conditions the birds are given. I am sure a proper free range chicken's egg is every bit as nutrious as a quail's egg from a battery cage.

When we were in Egypt in March 1998, we were looking at a quail project at one of the universities, and it was mentioned that quails can be identified by their eggs. I wasn't sure about this at the time, and wondered if I had lost something during the translation. In fact each quail has its own signature on its egg, in size and background colour, but the pattern of spots or blotches does vary a little with each egg; so hence it is possible for the keen observer to obtain a lot of information about their quail without actually trap nesting them.

Watch out for double yolker eggs, normally twice the size of a 10gm egg, and also for ones with a cloudy colour, as these eggs normally don't hatch well. There is a breed of coturnix quail that lays pale or almost white eggs, but these are seldom seen. You will notice that quail eggs alter with the seasons: in spring and autumn they are pale, but in winter they become darker.

INCUBATION

The incubation period for the domestic quail is 17/18 days. There are two main differences between the incubation and hatching of quail eggs and those of domestic fowl: quail eggs require a lower temperature, 37.5 degrees C or just under 100 degrees F, and lower humidity. Most quail eggs are incubated artificially, but this does not mean they cannot be hatched under a quail or a bantam. Most quail hens are not given the surroundings to make a nest and sit on their eggs, and very few choose to do so, but occasionally there will be the odd hen bird in an aviary or extended outdoor run that will sit, and they make good mothers too. Always remove the male bird as they tend to kill the chicks.

Quail eggs can be hatched under a bantam and even a pigeon, something I have never done, although I have had plenty of experience with partridge eggs. You would have to choose your bantam broody carefully, preferably a clean leg variety, and one with a little experience. Clean legs would be essential, as I have seen many a partridge chick tangled up in the leg feathers of a Silkie. Once past the 5 day mark, the chicks should romp away,

Back to incubators. As each incubator and location are different, incubation on a small scale can be a hit and miss affair in the beginning, but eventually with the aid of records you should be able to adjust your machine to obtain good hatches; (commercially quail eggs hatch between 93% and 97%). But then, just when you think you have cracked it, there can also be seasonal adjustments to make because eggs incubate and hatch differently in March from how they do in July. More information on incubation can be found in 'Incubation at Home', also in the Gold Cockerel Series. Just a word on humidity: you may require a little water during hatching in the early part of the season, but none at all in the summer, although again, I can only stress, this is not necessarily the norm. Certainly no water should be added during the first three days while the germ is forming.

It is best to use an automatic turning machine, as quail eggs are fiddly things to handle, and you must turn them at least three times a day. The large commercial incubators turn the eggs every hour.

Once the chicks start to pip, let them all hatch and leave them in the incubator for 24 hours. In large commercial outfits the eggs are moved into the hatcher two days before hatching. The trays in the hatcher are specially designed for quail, with sides and sometimes tops. Dividers are placed in the hatching trays so that separate blood lines can be kept apart. Quail chicks hatch out quite quickly, normally all within a

Quail eggs loaded into a commercial incubator. Note they are loaded point down, and the egg tray tilts backwards or forwards, every hour.

Quail chicks just hatched, note the size compared with a match box.

49

six hour period, and they are one of the few breeds to 'talk' to each other while in the shell to synchronise their hatch, which stems from the wild forebears who needed to leave the nest altogether and as quickly as possible. When eggs from two separate incubators are put in one hatcher, the two lots of eggs will hatch independently and at slightly different times, but there can also be an element of temperature difference in each incubator.

Temperature fluctuations in the incubator can lead to curly toes in new born chicks. While they are hatching avoid opening the hatcher / incubator, as this only leads to loss of temperature and humidity. Do not be concerned about feeding or watering the chicks during the first 24 hours, as they have a built in food reserve in the form of their yolk sac which is absorbed just before hatching.

CANDLING

It is possible to candle quail eggs despite the spots or blotches on the shell, see 'Incubation at Home', but most quail farmers and all the large commercial farms don't bother. As long as you are not breeding too closely, you know that you are going to achieve a 93-97% hatch. Candling is part of the school project on quail.

Be careful not to overheat them when candling these small eggs.

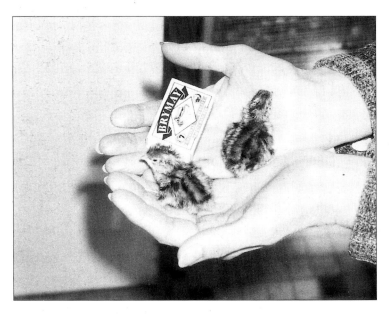

Newly hatched quail chicks

50

Hatching quail chicks in different compartments for recording and keeping different blood lines or colours.

BROODERS - These come in all shapes and sizes and are generally made of wood. The criteria for building a brooder are visibility, accessibility and economy.

Before building a brooder, decide which heat source you are going to use. These are available in several forms: electric bulbs (red), electric ceramic bulbs (heat but no light), electric elements and gas lamps. The electric bulbs and elements are the cheapest to run, the ceramic bulbs are next, with the lowest wattage of 100 watts, and gas is not really an option unless your production is large, and/or you have problems with electricity supplies, ie power cuts or non-availability of supply.

The system favoured by most small to medium producers, is the thermostatically controlled bulb heater, not just one but two, in case one fails. This will give a constant regulated heat and light by which the birds can see to eat and drink. These bulbs must be red or blue, but certainly not white or clear.

The broody box can be quite a simple affair, a 4' x 4' x 1' (122 x 122 x 30 cms) high box with a removable lid. The lid needs careful planning because you need to be able to clean the chicks, water them and feed them with the minimum of stress, and yet be able to observe all around the brooder and also allow for ventilation. When the chicks are small they require heat of 100 degrees but as they grow the temperature needs to be reduced (thermostatically) and ventilation and natural day light need to be increased.

The floor, sides and lid are best made of plyboard. A 4' x 4' x 1' high box (122 x 122 x 30cms) will accommodate 300 day old quail chicks. The illustration shows an all purpose brooder which I have used for years, with a heat lamp at one end. This brooder can be used for quail, ducklings and hen chicks, the size is 36" x 14" x 12" (92 x 36 x 30 cms) and will accommodate about 65 day old chicks. Remember that quail chicks grow rapidly, so space must be allowed for them. 50 chicks in this brooder would be fine until they are 3 weeks old. The floor can be covered with about 1" (2.5 cms) of 'dust free' shavings or corrugated cardboard. For a brooder 4' x 4' x 1', cut the cardboard into 2 strips 2' wide and put several strips on top of each other. Do not use newspaper or anything smooth as this will lead to slayed legs in the chicks; if you can't get hold of dust free shavings or corrugated paper, towelling will do.

Drinkers and feeders are made of plastic and are quite inexpensive, so it is better to buy the correct things rather than mess about with pans and saucers, see illustrations of both. You may have to put marbles or stones in the more open type of chick

A home-made brooder which can be used for quail as well as other species of ducks, hens and pheasants. The size is 36" x 15" wide x 12" high (lowest) (92x38x30.5cms). The brooder lamp can be raised or lowered and the plastic net covered lid over the run area prevents the quail chicks from jumping out.

Wooden brooders with two week old quail at the Agricultural College, Cairo.

Young quail 3 weeks old in a home made brooder, showing the heat bulb and plastic quail feeder.

drinkers for the first week, to stop any chicks from drowning. Young quail chicks are very thirsty, so it is most important to check the water levels frequently. If any chicks do become wet, dry them with a hair dryer.

Assuming the quail chicks have hatched and have been in the incubator for 24 hours it is now time to move them over to the brooder which should have been on for 2-3 hours beforehand to warm up. Sprinkle some food on the corrugated cardboard, and put in the feeders and drinkers with their marbles. Using a shoe box or something similar, move the chicks across to the brooder. Great care should be taken to ensure that the chick crumbs are small enough; quail chicks have starved many a time because the food was too large for them to take up.

As you can imagine the baby quail quickly become rather messy, hence the strips of corrugated cardboard. Remove the drinkers, and gently roll up the cardboard from one end, trying to stress the chicks as little as possible. Replace the drinkers and add the feeders after the second day. 4 or 5 layers of cardboard should last a week, and after that you can use shavings. The first week requires a lot of management, and thus it is important to have a lid on the brooder which is very easy to use.

By the second week the temperature needs to come down to 80 degrees F but bring it down slowly. Add more drinkers and feeders. Cleaning out is normally every two or three days but the main problem to avoid is toe balls on the chicks. These are hard lumps of manure which form around the toe nail on the foot. They can be picked off gently, but it is a time consuming job, and best to try and avoid this condition in the first place. See page 69 also. Remember that the chicks can flutter quite well as their wing feathers are growing fast.

At this stage, there can be a problem with feather pecking. This normally occurs when the brooder is too hot, the birds begin to sweat, and then start to peck at the feathers just above the tail. Losses can be considerable, so act fast and remove the pecked chicks, spray them with gentian violet and place them in a separate brooder. Check the heat in the main brooder, and cross your fingers that the feather pecking has stopped. Obviously you are keeping an eye on the food and water a minimum of twice a day and will continue this through the rearing cycle.

Into week 3, slowly reduce the temperature of the heat lamp or source to 70^0. Of course this will depend on the temperature outside, and in hot countries you may be able to turn the heat off for several hours during the day. The chicks can eat a large crumb food now and naturally will need to be cleaned out regularly.

Week 4. The chicks are feathered up and able to fly. Move them from the brooder box to the grower pens. Maintain a low light day and night, but turn the heat off or down during the day, (some electric lamps have adjustable heat variations); change the feeders and drinkers up to adult bird size. A low light is important to stop the chicks from cramming into the corners at night and suffocating. Quail chicks hate temperature fluctuations and will stress as a result. A word of warning when you move the chicks into a larger pen: ensure that the same coloured bulb is used, ie red or blue; never mix the colours as this will stress the birds. Now is the time also to start to sort out the cocks from the hens. Remove the cock birds to another shed, out of sight and sound of the hens. This way they will continue to grow without fighting. Care should be taken with coloured birds: if you are breeding coloured quail this is fine but if you have the odd white bird in your flock, it could well be set on by the others.

Week 5. The growers are getting larger and larger, and it is time to worm them with Flubenvet. This clears all the worms out of them so they are able to go forward to week 6, and start to lay without a withdrawal period. Flubenvet is added as a fine white powder to the food for seven days, and eggs should not be eaten during the week of administration.

Week 6. Turn the heating off at night but leave a low light on.

5 weeks old quail with wall mounted food hopper.

56

*A dual purpose and a meat quail
(both females)*

Melanistic Tuxedo

Golden Tuxedo

Tricolour Tuxedo

Although different colour variations are appearing in the Coturnix quail, there has not been any international standard on the name of these colour variations.

British White

Italian laying quail male & female

Light Pharaoh

Pharoah female

Golden

British Range

Pharoah male

Dark British Range

Red Tuxedo

Black Backed White

Melanistic

Fawn Backed White

COMMERCIAL BROODING AND REARING

The day old quail chicks arrive from the hatchery in plastic trays which are stackable for easy transport. Many quail farms in Europe only rear the young birds and do not breed quail or incubate their eggs. The chicks are taken into the brooder/rearing house, which is a large insulated building, pre-heated to 35 degrees C. There is a good bed of 'dust free' shavings about 3" - 4" (8-10 cms) deep on the floor, gas or electric brooder lamps overhead, automatic chick floor drinkers and plastic food pans on the floor.

The chicks are emptied out into a round hardboard or plastic corrugated board enclosure in the brooder room, to stop them from wandering away from the food and water. There is natural day light as well and subdued night light from 25 watt bulbs to stop the chicks from smothering each other. The round surround prevents them from finding a corner in which to pile up at night or in moments of stress. Most losses occur in the first five days while the chicks are still dependent on their yolk sac reserve and before they start feeding properly.

After a week to ten days, the plastic floor food pans and chick drinkers are gradually replaced with automatically fed floor hoppers and automatic hanging drinkers.

Quail rearing on a large scale. The plastic walling is to hold the 3 day old quail under the gas heater and near the food and water. The enclosed area is gradually increased every day.

I was interested to see this system work well in France, knowing the problems of quail chicks ingesting shavings instead of their food, but I believe this may be to do with the quality of the shavings (not 'dust free') and the food, which I discuss elsewhere. Because the brooder/rearing room was evenly heated, the chicks were well spread out and not huddled under each brooder lamp. This made walking around them quite difficult but you soon learned to gently kick them aside in order to put a foot down.

After the 3rd day, the heating is gradually reduced until it is turned off when the youngsters are feathered up at 3 weeks. Reduction is about 1 degree C a day from the 3rd day.

Lighting and ventilation are crucial under this regime. Light is reduced to a gloom as the birds mature. By week 5 the young adults have only a few hours of light, not enough for the males to mature fully and become sexually active. Even so, some males do mature and some femals start to lay, but by restricting the light one is normally able to hold back the sexuality of the birds and increase their body weight. It is possible to hold them 'in suspense' until week 12 after which they are fully adult and become too stringy to eat. Most breeders will slaughter their birds at 6 weeks because it is uneconomic to hold them over for much longer. In France the quail is now considered a "festive" bird (like turkeys in the UK), so there is a certain amount of holding over of the birds prior to Christmas and Easter,

Some people move their 3 week old quail into rearing cages, a typical size being 115 x 155 cm (45" x 61") with 90-100 birds per cage; but this system is gradually being phased out as the open plan or floor system is gaining ground, mainly because of equipment and labour costs.

The density of quails using the floor system is 80 chicks per square meter, with mortality between 6-8%. The shed is never cleaned out during the rearing period, but at the end of this time, the drinkers and hoppers are taken out and the floor litter is removed with a small tractor or Bobcat.

Most quail sheds have their own side and roof ventilation systems. Some use extractor fans, which are triggered by electric thermostats in order to keep a pleasant 65 - 70 degrees F. When you have 10,000 near adult quail in a building of 12 m x 10 m, the body heat produced by them can be considerable.

Transition stage, the floor feeding pans being phased out, and the young quail starting to use the food hoppers.

Automatic food hoppers, made in fibre glass especially for quail.

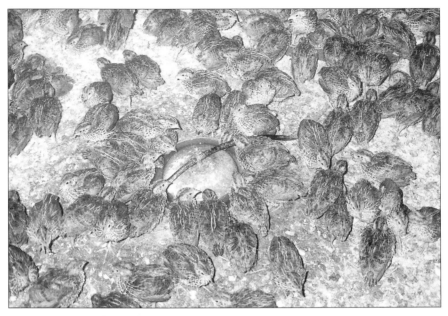

Young 3 week old quail using an automatic chick drinker.

Here is a round food pan which should have been replaced by a food hopper. These 3 week old quail are using the pan to eat and dust in.

60

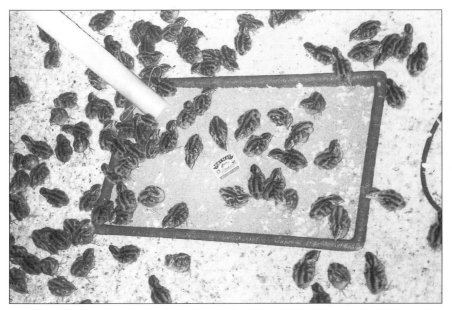

Two systems for feeding young quail. The top photograph shows a tray of food which is automatically filled. This is used on chicks up to 10 days old. Below, older birds about 3 weeks old feed from a hopper which is filled manually from a bucket or paper sack of food.

KILLING, PLUCKING AND FINISHING

This chapter is really in two sections. The first section is about home consumption and the second section concerns commercial systems. The two sections differ because of all the rules and regulations regarding the slaughter and sale of meat.

Killing time - Withdraw the food the night before, leaving the water drinkers working, then kill the next morning. This way the birds' crops will be empty when you come to prepare them.

Home consumption. Killing - The best and cleanest method of killing quail, from young to adult birds, is with 'Semark pliers'. These are flat bladed pliers, which are placed around the neck behind the head, and the handles are then squeezed shut. This severs the vertebrae and breaks the nerve cortex. There is no blood or mess, although the bird continues to wriggle and flap for about 20 seconds.

To pluck or not to pluck - Quail feathers do come off very easily, as you will realise from handling live birds, but it is quite time consuming plucking quail, particularly if you have to do several. As with snipe or woodcock, it is easier to skin them. Just nick the skin inside the leg joints and gently separate it from the flesh. When you come to the wings, skin up to the first joint and then cut off the rest. Chop off the head and neck at the base of the neck.

Gutting - Some people gut quail, others don't. It is a matter of how strong you want the birds to taste. If you gut them, slit the skin and the flesh gently with a sharp knife taking care not to puncture the intestines, about 1 1/2" (3.5 cms) above the vent (the bird facing breast up) then ladle out the insides with a spoon. This is best done on a newspaper so as soon as you are finished, the innards can be rolled up and disposed of. Some people like to hold back the liver and heart for making stock. Just wash out the cavity with cold water and dry the bird with paper towelling; you are now ready to cook or deep freeze it.

If you have skinned the bird and are going to roast it, take care that the flesh does not dry out. To overcome this lay a piece of bacon on either breast. Skinned quail are excellent in a casserole.

Quail destined for the table.

Quail food and catching nets.

COMMERCIAL SLAUGHTERING - The methods will vary slightly from abattoir to abattoir, country to country, but the emphasis is on getting the maximum volume of birds through as there may be ten or more people on the line. Most of the people on the line are women, with men doing the heavier work at either end. I will run through the process of a typical abattoir I saw in France.

Quail killer. The crate of quail slides into this metal box, the door closes tightly, and the air is extracted. This takes about 30 seconds, and the quail lose consciousness immediately.

The quail are first caught up in the rearing houses using fishing landing nets; approximately 30 birds are put into each plastic crate. The food is withdrawn the night before, and the birds caught first thing the next morning as some abattoirs start early to avoid the mid-day heat.

The crates arrive at the abattoir, and are placed onto a roller conveyor where they await the killing box which is a metal container with a sealed door. Once a crate is inside the killing box the air is quickly extracted, and after 30 seconds the crate is removed with the dead quail inside. The quail are taken out, fixed onto overhead leg shackles, and then moved on to the conveyor chain to the rough plucking department. Here the birds are held against dry plucking machines: these are metal disks which rotate concentrically, pulling out the feathers which are sucked away by a vacuum. They are held in this position for about 10 seconds, long enough to remove the long feathers on the wings and tail and those up and down the body: very fast and noisy work. The nearly naked birds then move on to the hot wax bath where they are immersed up to the neck, after which they are conveyed through two cold baths of water to solidify the wax. They then move on to the wax peelers. There are normally six women, three either side of a rubber conveyor belt, who undress or dewax the bodies. This is just like taking off a waistcoat, and all the feathers and fluff adhere to the wax which falls onto the conveyor belt and goes off to the wax melter. Once the wax is hot, the feathers are filtered off and the wax is recycled back into the system.

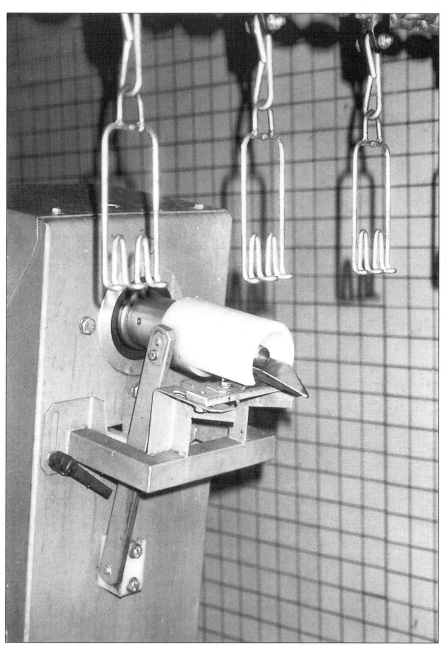

A degutter or eviscerator. The round metal tube inside the white plastic guard enters the abdominal cavity and sucks out the innards under pressure.

The rough plucking machine which removes the wing and tail feathers and most of the body feathers before the wax bath.

The bodies are moved next to the gutting machine, (often called an eviscerator). The bird is presented to the machine vent first; it is cut and a metal tube enters the cavity, sucking out everything from the inside under pressure, very quick and clean. The body moves on to have the legs and feet cut off by two small circular saws, and is then removed from the shackles to be graded, weighed and packed.

Obviously there are many more details involved with this system, concerning timing, temperatures and techniques but anyone going into this would obviously consult the manufacturers of this very specialised equipment.

In France, the French producers often leave the head and neck complete with feathers on the bird, as a clever marketing ploy. It helps with the net weight, as in general the bird loses a quarter of its total weight after preparation for the table, and the French also feel that it looks more natural.

Packing is a subject that could be endlessly explored, but a lot depends on the destination. At the abbatoir we visited, birds were packed in twos, fours, eights, etc, placed on a polystyrene tray, shrunk wrapped and labelled. This was mainly for the supermarkets, while birds destined for butchers and the catering trade were bulk packed in cardboard boxes.

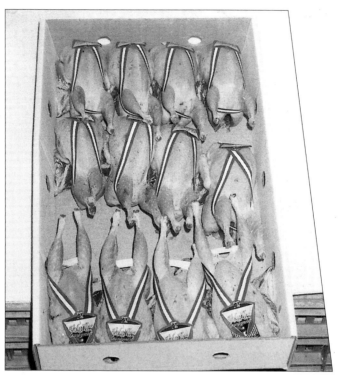

Quail freshly processed and ready to be sent out to a butcher.

Quail off the finishing line, graded and ready to be custom packed.

67

DISEASES & AILMENTS

Quail are like pheasants and poultry, they can suffer from most of the diseases that pheasants and poultry can have, but in fact they rarely do. Most of the problems quail experience are connected with their food and surroundings. They are normally reared intensively indoors and because they are reared on clean shavings and their life span is only 6 weeks for meat quail and 8 months for egg laying quail, there is little time for any chronic diseases to set in. Having said that, there are certain diseases and ailments which quail are prone to and we will look at those in relation to three different methods of keeping them.

QUAIL ON THE FLOOR - The main problems here are respiratory (aspergillosis), coccidiosis, worms, stress, feather pecking and toe balling.

When a quail is sick, its feathers are puffed up and it looks sleepy and cold and drags itself around; so what is it suffering from?

Respiratory - This is fairly easy to see, as the bird is trying to catch its breath: this is usually aspergillosis and is caused by a fungus which is found everywhere, often associated with damp litter of shavings or straw. It can also be caused by stale food. The birds need to be moved to a clean dry area. If the house is dusty, that can also cause respiratory problems.

Coccidiosis - The birds look off colour, often with milky white or runny droppings. This is frequently seen in floor reared birds, the coccidiae rising up from the ground even off concrete. A coccidiostat in the water or food will clear this up.

Worms - Not often seen, but the symptoms are the birds or bird going light, and feeling like a skeleton under the feathers. Worm that batch of quail with Flubenvet in the food for 7 days, and clean out the birds at the end of that period.

Stress - Is one of the most difficult problems to put your finger on. It can have a multitude of causes, and here are some: lack of food and/or water, too cold, too hot, too draughty, too crowded, change in food or protein, not enough drinkers or food hoppers, too much light, sudden noises, vermin (rats or mice). The symptoms can vary from feather pecking, dead birds and birds piled in corners, to a drop in egg production, and birds not growing.

Feather Pecking - Is a symptom of stress, but can break out even under the right conditions, very often caused by a change in foodstuff and protein levels. Just rub the pecked area with Vick vapour rub, and return the birds to their pen. In persistent cases, plastic bits or beak clips (or as the Americans fabulously call them 'Aggression restrainers') can be used, although nobody bothers in the commercial world.

Toe Balling - This only comes about from quail padding around in dirty litter. Droppings ball up around the toes and feet of the bird and harden. These need to be picked off with the greatest of care, otherwise you will pull the toe nail out as well, and they bleed profusely.

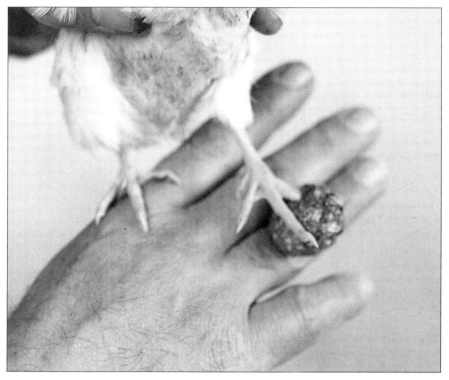

A problem with dirty litter in floor pens. This is a toe ball of quail droppings and shavings which has hardened, and is difficult to take off without injuring the bird.

QUAIL IN CAGES - Quail in cages are mainly laying birds and they suffer from respiratory problems, stress, feather pecking and back scratching, prolapse and beak and claw deformities.

The respiratory problems are mainly caused by dust, so it is important to have an efficient extraction system, as well as a good hygiene programme. In reality, layer cages are only cleaned at the end of a cycle, although the droppings are removed weekly, so laying houses can become very dusty from the food and feather particles.

We have already dealt with stress and feather pecking.

Back Scratching - This is caused by the cock birds mounting the females. Often the hen birds can be devoid of feathers on the head and neck and down the back. There is little that can be done, although reducing the number of cocks can help.

Prolapse - This is when the oviduct protrudes out of the vent, and happens mainly in the first two weeks of lay. There is always a small percentage of birds that lay double yolkers which can cause prolapse. Often the other birds will get to the affected one before you do, and pick its insides clean, leaving a hollow carcase. There is no cure for prolapse so dispatch any birds with this condition.

Beak & Claw deformities - This is quite common in birds kept in cages. Because they are in unnatural surroundings the beaks and toe nails can grow very long, but can be trimmed with nail clippers.

QUAIL IN AVIARIES - The susceptibility of quail to disease here is more because the aviary is exposed to the outside world and possible contact with other birds, including wild ones. Also, quail kept in aviaries tend to live longer, and so there is

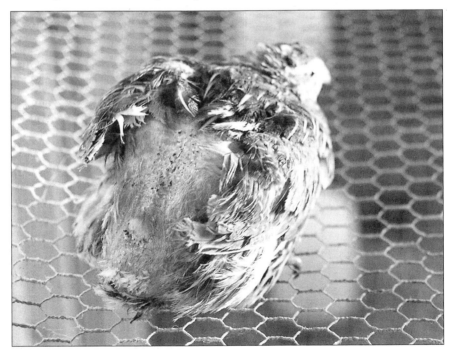

This is the result of male or cock activity on this hen bird, caused by the male trying to tread the female. The back of the head is also affected.

70

a slight possibility of catching diseases and ailments associated with this, such as avian tuberculosis and leucosis, but it would be very rare, and normally the problems to look out for are as follows: worms, coccidiosis, blackhead, mycoplasma, scaley leg, fleas and mites and feather pecking.

We have already dealt with worms, coccidiosis and feather pecking.

Blackhead - This disease is caused by a parasite called Histomonas meleagridis which is spread by a roundworm called Heterakis. The eggs of this roundworm can be carried by earthworms which the quail can pick up in the aviary, so there is a possibility of the quail contracting this disease. The birds look hunched up and cold, and have bright yellow droppings. This can be cured by adding dimetridazole to the drinking water.

Mycoplasma - This disease is becoming more and more common. It is an infection of the air passages in the head, neck and lungs, causing rattling when the birds breathe, discharge from the eyes and nostrils, sometimes rather smelly, and the birds being rather dull and lethargic. Treatment in the water with Terramycin or soluble Tylan normally clears this up. There is a possibility that soluble Tylan may be taken off the market in the future but it is still available at the moment.

Scaley Leg - This is caused by a mite which burrows under the scales of the legs and feet, lifting the scales and in severe cases causing a crusty white 'bark' on the legs and feet. This can be cured by dipping the leg and foot in surgical spirit once a week for 5 weeks. It takes about 6-8 months to clear up.

Fleas & Mites - These are normally carried by wild birds. The quail look dull and lethargic, often scratching themselves. Several dustings of pyrethrum powder will normally clear this up.

Just a couple of notes about aviary quail: before you first place them in the aviary, clip one wing so they can't fly. This way there is no danger of them 'head banging' and causing themselves an injury. Secondly, aviary quail are very prone to predation by rats so make sure your aviary is vermin proof.

Moult - Quail moult like other birds. The main moult you will notice is at 4 to 5 weeks, just before coming into lay or sexual maturity. After that they moult like wild birds, a few feathers at a time over a 10 week period. Quail can be stress moulted by altering light patterns or withdrawing food or water, but this is not practiced as the life span of a commercial quail is so short.

A commercial quail's life span is normally 8 - 12 months after which they are culled, but this does not mean that they are useless after this period. From the commercial view point, that is the length of maximum productivity; after that egg production drops, and so does the number of birds, by as much as 50%. After laying for a year in cages, quail can be used as aviary birds; they will not lay in the winter but will live happily for 3-4 years.

QUAIL IN THE SCHOOL - We are often asked what diseases quail can transmit to children. The answer is none, but we are sure that someone will invent something! There are however two points to look out for: first, some children might have an allergy to feather dust, and secondly if you have been handling the birds, it is a wise precaution to wash your hands afterwards. Otherwise quail are quite safe.

Just a word about fertility in quail. While most books categorically state that you require 1 cock to 3 hens, results can be just as good with 1 cock to 6 hens. There is less scratching of the hens' backs and less fighting among the males. One point to remember, cocks are not sexually active when they are cold, so you must maintain an even warm temperature in the breeding area.

There are two quail diseases in the U.S.A., Quail disease and Quail cholera, which fortunately we don't see in England. For further information about these contact your local avian vet.

ECONOMICS

Although the economics may look interesting, there are certain details that I am unable to provide, namely the capital cost of setting up and the day to day running costs. It is important to start small and build up which can be done quite easily because of the fast reproductive cycle of quail. Here are some facts and figures to help you with your calculations.

The price of a laying hen quail can vary from 75p to £3.00 for coloured or pet quail. Males are less, 50p each.

Each hen will lay (if an egg laying strain) about 300 eggs per annum ie 25 dozen.

These eggs, if sold wholesale about £1.00 per dozen = £25 per annum, if sold retail about £1.90 per dozen = £47.50 per annum, hatching eggs are sold for £2.00 - £2.50 per dozen = £50 per annum.

A laying hen quail eats about 5oz per week x 52 weeks = 16 1/4lbs of food (7.38 kilos) per annum. A growing quail up to 40 days eats 17ozs (480 gms) of food, approximately £0.14 per bird. A 55lb bag (25 kilo) of quail food costs approximately £7.50, so each quail eats about £2.30 per annum. But, here are the unknown costs: water, electricity, cage costs, drinkers, feeders, building costs or rental costs, marketing, labelling, packaging, travelling costs and your labour and time.

Quail for zoo food fetch between 40p - 75p each. Again I haven't put in the unknown costs.

Before you rush into quail and think you are going to make your fortune, you must find your market. In 1995 the quail produce market in France was worth 250 million francs.

THE LAW - Common quail (wild corturnix quail) are a schedule 4 bird under the Wildlife and Countryside Act 1981, and as such must be registered with the Department of Environment (D.O.E). One reason for this is that quail fighting is a regular event within certain immigrant communities. The act does not apply to domestic quail or their keeping

MARKETING

If you are intending to start a quail business, it is essential that you have some experience with quail and their management, if only to be able to answer all the questions you will be asked. This experience can be obtained on a small scale, and it is quite easy to expand the business quickly, bearing in mind a quail egg can produce a laying hen bird within 10 weeks, so in theory you can multiply the size of your business 5 times in a year.

Start small and try different markets, and then as the business takes off, specialise in the area where you find the best returns, or that best suits your life style.

One of the problems with selling quail produce, be it eggs or meat, is that it is dogged by rules and regulations; this is why there are more small producers than large ones in this country and the large producers tend to import produce from France. The economics look good and there is a market out there, despite the reserved British palate. The largest share of the market is in the south east of England, but let us look at the possibilities and the different quail products and markets.

FRESH QUAIL EGGS - The sale of quail eggs carries few restrictions and there is a good market for these in England. The main outlets are supermarkets,

Quail eggs packed and being put in an outer cardboard box ready for despatch to a supermarket.

74

delicatessens, Farm Shops, butchers, pubs, restaurants, etc. It is easier for the beginner to deal with smaller shops than supermarkets, who are on the whole very fickle and tend to hike the retail price. Quail eggs are sold in clear plastic containers to show their attractive colouring, and normally in 10s, 12s or 24s. These containers must carry the name and address of the producer, packing station number (this you obtain from your local MAFF Egg Officer together with all the rules and regulations regarding the sale of eggs) and the date when packed. Prices can vary according to area, location and shop, and it can be a seasonal market, with Christmas being very busy.

COOKED QUAIL EGGS - This is the product that is going to take off in the future. Although there are a few more rules and regulations, these are mainly common sense and practical; you would have to consult with your local authorities. Some people find a good market for quail eggs in brine or vinegar, in delicatessens, pubs, Farm Shops, etc but there could well be a market for cooked quail eggs in the form of mini scotch eggs and mini veal and egg pies sold in plastic packets of 6 or a dozen; there could also be all sorts of cocktail stick ideas, cooked eggs for decorating salads, hors oeuvres, curried quail eggs, garnishes etc. These are all interesting possibilities for the future, and the advantage is that there is no seasonal factor to take into consideration.

QUAIL FOR ZOO FOOD - Although I have headed this section as zoo food it mainly concerns raptors: there is a good market for food for hawks, falcons, eagles and owls, but also for certain reptiles and snakes, and small carnivores such as linx, civets, foxes etc. This market is mostly supplied by frozen birds; some people want young ones, some young adults, and others will cope with old hens from the laying pens. There is even a project to feed up old hens on vitamin enriched food, to enhance their nutritional value. The birds can be posted or sent by carrier in small or large quantities, and are advertised in specialist magazines and papers, like Cage and Aviary Birds.

The quail are killed with Semark Pliers, and allowed to cool. They can be packed in plastic bags in 10s, 50s, 100s etc according to the customers' requirements, and can be collected freshly killed as well. You should stick a label on the bag with your name and telephone number (address is optional) together with the date of slaughter, and keep a record of all transactions. Raptors etc obviously have to be fed all the year round, thus there is a steady market with an increase in the summer months when the raptors are breeding.

REARING QUAIL - There is a growing market albeit quite seasonal, for rearing quail, mainly egg laying ones. These are provided as either hatching eggs or day old chicks and are reared and sold back, payment being made on a per capita basis. The prices are not terrific, inter producer markets never are, but it can help to spread the labour costs and also help with the feed bills. This is where close contacts within the trade are essential.

QUAIL FOR PET SHOPS - There is a growing market for coloured quail, mainly for people who have avairies. As you can see from the photographs, the colour variations coming through are very interesting. These birds are poor egg layers as a rule, perhaps only producing 80 eggs per annum. You will have to contact your local pet shops and garden centres to build up a round and a service which can also include other quail products. With the absence of a national carrier to handle live quail, the market has to be regional rather than national, but this of course does not apply in the U.S.A.

QUAIL FOR LABORATORIES - This market can cover hatching eggs as well as young adult birds. It is not a market I have looked into, but I know it exists and it would need to be investigated further. Quail make excellent laboratory birds because of their size, their fast reproduction, their light sensitivity, their ability to be sexed fairly early in life, their egg production, etc.

QUAIL FOR MEAT - Until recently the British have been very reserved about eating quail, although they have probably eaten it in game pate without realising it, as most of the old laying hens in France are made into pate. However, with the advent of mass tourism, people are being exposed to different foods, and demand for quail meat is increasing. In France and Italy quail are now the food that people buy for preference at Christmas and Easter; it is still seen as a luxury product, but more and more people are turning away from traditional festive meats such as lamb, because they don't want to spend the hours in the kitchen, nor have they such large families to feed. Quail is a simple, easy and quick alternative (6 minutes in the microwave and 20 minutes in the oven) and without the leftovers which can hang about for days over the festive period.

The problem lies with the rules and regulations which are daunting. My thoughts on this are: if you are already producing and selling turkeys, cockerels, pheasants, etc and you are in the know about the requirements and are already set up, then it is simple enough to bolt on quail. You already have the approved buildings, the tiled floors and walls, the chill room, refrigeration, and hygiene controls etc, plus the market outlets as well.

Here, as with the sale of other birds, presentation is all important, together with labelling, recipes etc. Quail steals a march on pheasants and partridges, as it can be sold all the year round, rather like guinea fowl.

QUAIL FOR THE ETHNIC MARKET - There is a market for the sale of adult quail mainly to Indians and Pakistanis. They like to buy the birds live and kill them at home. Again it is a matter of contacting the right people, often to be seen at local poultry markets around the UK.

RECIPES - I am not a cook, although I enjoy being a menace in the kitchen, so I haven't padded this book out with endless recipes, as these can be found elsewhere, written up by far more capable and experienced people than myself!

A pack of 20 quails eggs as sold in France. Note the woodland scene to give an impression of a natural product.

SCHOOL PROJECT

Quail make a colourful but sometimes noisy project in the school. Because of their size, they are easy to accomodate, and they have certain characteristics which make them more worthwhile than chickens or ducks.

Before such a project is planned, thought must be given as to their management at the weekends, half terms and holidays. The best system is to have a set of cages at school and a set of cages at home, and take the birds home at weekends etc, in a cardboard box. A garden shed or conservatory works well to accomodate your quail but it must be somewhere safe from your cats and dogs. It is probably best also to have two sets of drinkers and feeders and food. It must be noted that the males have a shrill trisyllable call, so a suitable location in the school will be important in order to avoid disturbance. You may also need to give the birds additional lighting on a time switch to ensure that they lay eggs according to plan. Quail enjoy dust bathing so can be a little messy, a point to consider when deciding where to have your birds.

WHAT STUDIES CAN BE MADE - With laying quail (hens only) studies can be made of the amount of water and food consumed, the colour of eggs, the size and shape and the number of eggs. Each quail has her own colour and shape of egg, although sometimes this is difficult to follow.

It doesn't matter if you have no male quails as you can buy hatching eggs and incubate them. Studies can be made of the weight of the egg during the incubation period, and the development of the embryo; watch out also for the synchronised hatch.

When the chicks are hatched, a study can be made of daily weight gain, water and food intake, feather growth, the emergence of the different sexes and the first moult.

These are some of the most obvious studies that can be undertaken. At the end of the study period the birds can be returned to the source of supply, to a pet shop or perhaps adopted by some of the children from the school!

QUOTATIONS AND SAYINGS

Quail is derived from the medieval latin word Quaccula, probably of imitative origin.

Quail is sometimes written as quaail, quale, queel, queal, quill, quilly, quaill, quaylle, quayle, quaile, qwayle & qwyle.

John Dryden (1631-1700)
> *We loathe our manna, and we long for quails*

Rev. Patrick Bronte (1777-1861)
> *No quailing, Mrs Gaskell! No drawing back*

Wallace Stevens (1879-1955)
> *The quail whistle about us their spontaneous cries*

Geoffrey Chaucer (1340-1400) The Clerkes Tale
> *And thou shalt make him crouch as doth a quail*

Sir Noel Coward (1899-1973) Children of the Ritz
> *We say just how we want our quails done*
> *And then we go and have our nails done*

William Shakespeare (1564-1616) Anthony & Cleopatra
> *But when he meant to quail and shake the orb he was as rattling thunder*

A Midsummer Night's Dream
> *O fates, come, come, cut thread and thrum*
> *Quail, crush, conclude and quell*

III Henry VI
> *This may plant courage in their quailing breasts,*
> *For yet is hope of life and victory*

I Henry IV
> *There is no quailing now*

Anthony & Cleopatra
> *His quails ever beat me, in hoop'd, at odds*

Troilus & Cressida
An honest fellow enough, and one that loves quail

Cymbeline
My heart drops blood, and my false spirits quaile

Old Somerset saying circa 1874
I'm quailin' away vor want o' zummot to ait!.

French expression.
Chaud comme une caille: hot as a quail (someone who is rather ardent)

A French term of affection
Ma petite caille: my little quail

Quailery (caillere in French) a place where quails are kept to be fattened for eating.

Quail pipe. A pipe used to call quails

Quail pipe boots.
Boots resembling a quail pipe from the number of pleats and wrinkles

Lait caille (quailed milk). Means curdled milk as in cheese making

Palsgrave - *This mylke is quayled, eat none of it*

Batchelor's Orthoep. Anal. *The cream is said to be quailed when the butter begins to appear in the process of churning.*

Coaillier is Old French for 'to curdle'.
The English word coagulate (to clot or curdle) derives from this.

From Thomson's seasons:
A fresher gale begins to sweep the wood and stir the stream sweeping with shadowy gusts the fields of corn, while the quail clamours for its running mate.

From Ebenezer Elliott's Morning:
Harps the Quail amid the clover,
O'er the moon-fern whews the plover

The Sussex poet Hurdis, in The Village Curate.
> *So have I seen, the spaniel-hunted Quail with lowly wing*
> *Shear the smooth air: and so, too, have I heard*
> *that she can sweetly clamour, though compell'd*
> *to tread the lowly vale*

From Bishop Mant.
> *Less likely of your aim to fail*
> *If with loud call, the whistling quail*
> *attract you, 'mid the bladed wheat*
> *to spread the skillful snare and cheat*
> *with mimicks sounds his amorous ear,*
> *Intent the female's cry to hear.*
> *For now the vernal warmth invites*
> *From Afric's coast their northward flights*
> *And prompts to skim on nightly breeze*
> *Sicilian or Biscayan seas*

A Quail can mean in slang either a harlot or a courtesan; the origin is said to be the south east of England.

Motteuz (circa 17th century) *'with several coated quails,* (dandies dressed in their finery), *lac'd mutton, waggishly singing'*

A Quail was a term used for a girl student in American university slang.

Quail-pipe was slang for a woman's tongue. Circa 17-19th century.

1084 AD. - *I stood as stylle as a dased quayle*

W. Waterman - 1555 - *'quaill, and mallard are not but for the richer sorte*

CONVERSION TABLE (IMPERIAL TO METRIC)

	1 oz	=	28.36 gms	55lbs = 25 Kilos
	2 oz	=	56.72 gms	
	3 oz	=	85.08 gms	
1/4lb =	4 oz	=	113.44 gms	
	5 oz	=	141.80 gms	
	6 oz	=	170.16 gms	
	7 oz	=	198.52 gms	
1/2lb =	8 oz	=	226.88 gms	
	9 oz	=	255.24 gms	
	10 oz	=	283.60 gms	
	11 oz	=	311.96 gms	
3/4lb =	12 oz	=	340.32 gms	
	13 oz	=	368.68 gms	
	14 oz	=	397.04 gms	
	15 oz	=	425.40 gms	
1 lb =	16 oz	=	453.76 gms	
	17 oz	=	482.12 gms	
	18 oz	=	510.48 gms	500 gms = 1/2 kilo

In some of the photographs I have used a matchbox in order that you can see the scale or size of birds etc. This is the actual size of the match box.

LIST OF SUPPLIERS

SUPPLIERS OF QUAIL (ENGLAND)

Clarence Court Ltd, P.O. Box 2, Winchcombe, Glos, GL54 5YA.
Tel: 01 242 621 228
Fax: 01 242 620 145

Curfew Coturnix Quail, Buttons Hill, Southminster Road, Althorne, Essex, CM3 6EN
Tel: 01 621 741 923
Fax: 01 621 742 680

Game for Anything, Rosehayes, Clayhidon, Devon, EX15 3TT
Tel; 01 823 680092

E.F. & F.E. Reeves, Meadow View Quail, Church Lane, Whixhall, Shropshire, SY13 2NA
Tel: 01 948 880 300

Quail World, Glenydd Penrhiwllan, Llandysul, Dyfed, SA44 5NR
Tel: 01 559 370 105

SUPPLIERS OF QUAIL (FRANCE)

Vallee De La Vie, La Retiere, Mache Vendee, France, 85190
Tel: 51 55 72 62
Fax: 51 54 23 79

SUPPLIERS OF QUAIL AND QUAIL EQUIPMENT (U.S.A.)

Lyon Electric Inc, 2765 Main St, Chula Vista, California, 91911
Tel: 619 585 9900

Strombergs Unlimited, P.O. Box 400, Pine River, Minnesota, 56474
Tel: 218 587 2222
Fax: 218 587 4230

USEFUL PUBLICATIONS

Cage & Aviary Birds, I.P.C. Magazines, Kings Reach Tower, Stamford Street, London, SE1 9LS

Country Garden & Smallholding, Station Road, Newport, Saffron Walden, Essex, CB11 3PL

Smallholder, Long Street, Dursley, Glos, GL11 4LS

Game Bird & Conservationists Gazette, P.O. Box 171227, Salt Lake City, Utah, 84117, U.S.A.

SUPPLIERS OF QUAIL EQUIPMENT

Brinsea Products Ltd, Station Rd, Sandford, North Somerset, BS19 5RA
Tel: 01 934 823 039 - Incubators, Candlers, Brooders

Curfew Incubators, Buttons Hill, Southminster Road, Althorne, Essex, CM3 6EN
Tel: 01 621 741 923 - Incubators, Cages, Candlers, Brooders

Danro Ltd, Unit 68, Jaydon Industrial Estate, Station Road, Earl Shilton, Leicester, LE9 7GA
Tel: 01 455 847 061 - Quail Egg Boxes, Cartons & Designed Labels

Gamekeepa Feeds Ltd, Southerly Park, Binton, Stratford-Upon-Avon, Warwickshire,   CV37 9TU
Tel: 01 789 772 429 - Drinkers, Feeders, Incubators (send for catalogue)

Boddy & Ridewood, Eastfield Industrial Estate, Scarborough, N Yorkshire, YO11 3UY
Tel: 01 723 585 858 - Drinkers, Feeders, Equipment (send for catalogue)

QUAYLE (H. Coll). Argent, on a chevron sable, guttée-
d'eau, between three quails proper, in the centre chief point

a pellet, two swords points upwards of the first. **Mantling**
sable and argent. **Crest**—Upon a wreath of the colours,
upon a mount vert, a quail as in the arms, between two
bulrushes proper. **Motto**—"Qualis ero spero."

Quail used in a coat of arms by the Quayle family.

A diagram showing the bird world and the relative position
of Quail

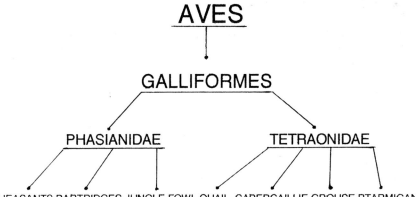

85

Modern Free Range   ISBN 0 947870 059
A practical guide for the small farmer on free range egg production and sales.   54pp. PB

Chickens at Home   ISBN 0 947870 07 5
Four methods of keeping chickens or bantams in the garden or smallholding plus practical
advice on feeding, watering, management, diseases, hatching and rearing.   56pp. PB

Bantams in Colour   ISBN 0 947870 00 8
The first colour photographic book of 40 breeds of Bantams and miniature fowl.
121 photos, 48pp. PB

British Large Fowl   ISBN 0 947870 11 3
The long awaited comprehensive history of 22 British Breeds plus 82 colour illustrations.
68pp. PB

Ducks and Geese at Home   ISBN 0 947870 09 1
A practical guide for keeping and rearing domestic waterfowl in small areas and free range.
56pp.PB

Domestic Duck and Geese in Colour   ISBN 0 947870 03 2
The first colour photographic guide to all breeds of domestic ducks and geese, over 80
pictures.   40pp.PB

Wildfowl at Home   ISBN 0 947870 08 3
A beginner's guide to keeping and breeding ornamental waterfowl written by Alan
Birkbeck and edited by M. Roberts. Charts for compatibility and breeding.
64pp. 38 colour plates. PB

Turkeys and Home   ISBN 0 947870 06 7
The first turkey book for small scale producers for 36 years. Practical sections on breeding,
housing, rearing, management, nutrition, diseases and exhibition.   64pp. colour plates. PB

Modern Vermin Control   ISBN 0 947870 04 0
A practical guide to the control of vermin for the smallholder, aviarist, farmer, poultry man
and game keeper.   56pp. PB

Incubation at Home   ISBN 0 947870 16 4
A detailed guide to working small incubators, and most of the problems encountered with
incubation.   56pp. colour plates. PB

Poultry House Construction   ISBN 0 947870 21 0
A DIY guide to building poultry houses and allied equipment.   92pp. PB

Poultry and Waterfowl Problems   ISBN 0 947870 26 1
A poultry keeper's and veterinary guide to diseases and complaints of hens, geese, ducks
and turkeys.   PB

Other Books in
The Gold Cockerel Series

Available from
DOMESTIC FOWL RESEARCH
Kennerleigh Nr. Crediton, Devon.
EX17 4RS England